Animals & Men

The 2013 Tasmania Expedition

Two different mystery cats in
Worcestershire; Fortean woodlice; Weird Weekend
2014; News, reviews and more

The Journal of the Centre for Fortean Zoology; Issue 52; February 2015

Contents

Typeset by Jonathan Downes,
Cover and Layout by SPiderKaT for CFZ Communications
Using Microsoft Word 2000, Microsoft Publisher 2000, Adobe Photoshop CS.
First published in Great Britain by CFZ Press

CFZ Press, Myrtle Cottage, Woolsery, Bideford, North Devon, EX39 5QR

© CFZ MMXIII

ISBN: 978-1-909488-26-7

Faculty of the Centre for Fortean Zoology

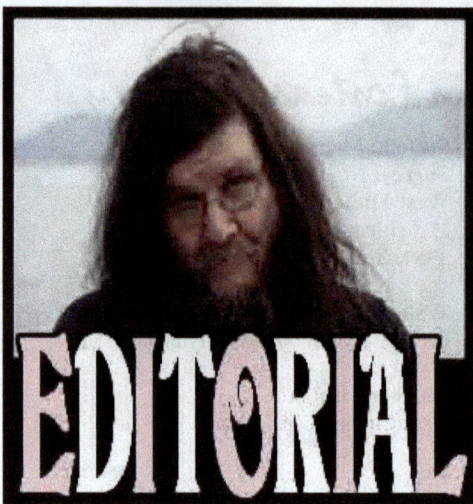

EDITORIAL

Dear Friends,

I apologise for it having taken such a ridiculous length of time between issues, but - to be honest (and I always try to be honest) - the world in general, and the cryptozoological community in particular, has changed dramatically in the last few days, and I have spent much of that time trying to work out how I could adapt the CFZ to meet the needs and challenges of this brave new world.

I said a lot of this in my Annual Report which was published on New Year's Eve, and as I don't want to repeat myself in this editorial, I have reproduced the relevant parts of said report later on in this magazine.

But as I said in the introduction to the report:

> "This one will be much shorter than usual and heralds the most radical shake-up of the Centre for Fortean Zoology this century.
>
> The truth is that times are changing, and we have to change with them. Like John Lennon famously said about The Beatles in 1969, we are in danger of becoming an "...ancient monument, and ancient monuments should be changed or scrapped"."

I find it hard to admit the uncomfortable truth, which is that after twenty years of running the Centre for Fortean Zoology, 2014 was the year that I very nearly closed it all down.

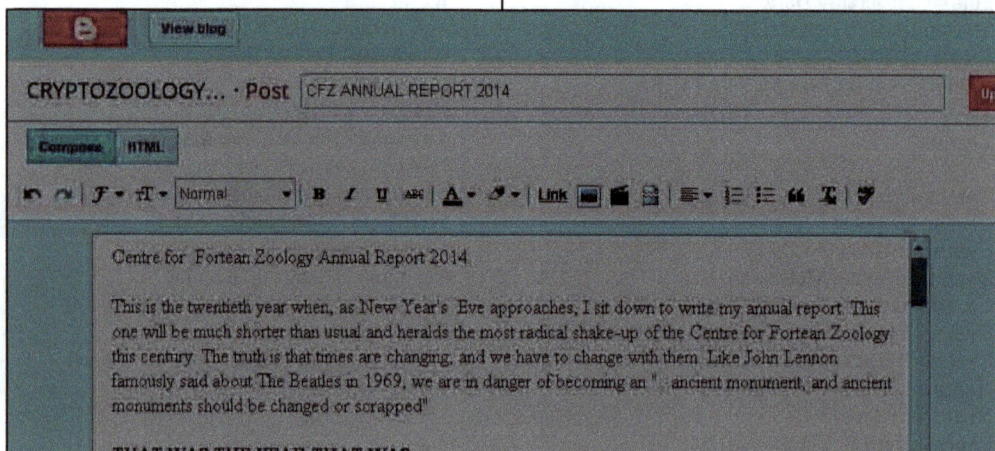

The Great Days of Zoology are not done!

But I didn't, and this issue of *Animals & Men* is the first fruits of the brave new CFZ.

The big questions that we have to ask ourselves are:

1.   What is Cryptozoology for?
2.   What is the CFZ for?

In recent years, as I wrote in my Annual Report, cryptozoology has become devalued in many people's eyes, as it has become the subject of a number of facile 'Reality TV' shows, and 'documentaries' which have largely been a lot of nonsense.

In the past few years we have seen one reasonably well-regarded TV show claim that the Loch Ness Monster is a Greenland shark (*Somniosus microcephalus*) which is not only a deep water species, which is found further into the Arctic than any other elasmobranch, but is NOT found in

Did Jeremy Catch the Elusive Monster of Loch Ness Lore?!

fresh water.

We have also seen a TV show, on which I - to my embarrassment - appeared, which claims that the 'Beast of Dartmoor', a name given by the show's producers to a video of what is obviously a wild boar (probably one of the ones liberated by

Animal Rights protestors near Fernworthy reservoir, Manaton a few years ago) was a hybrid between a wild boar, or maybe a Dartmoor pony and a lion!

This, despite both me and Danny Bamping (batting for the same side for once) explaining in tortuous detail what it actually was. I don't know about Danny, but I was filmed for an hour and a half, and they cherrypicked a few quotes from me that made it look as if I was supporting their daft theory.

But this happens a lot. Back in 1998 when Graham and I were in Puerto Rico filming for UK C4, some locals showed us what were obviously worm casts on the dusty ground. I was filmed kneeling by them (those were the days, when I could actually kneel) and saying something like: "These are wormcasts from some species of earthworm. But such is the level of the local people's belief that they are convinced they are the droppings of a chupacabras". You will not be surprised to hear that only the last five or six words made it to the final cut.

I have spoken at length elsewhere about the current rash of bigfoot-related Reality TV shows, so I won't repeat myself. *Shooting Bigfoot*, a documentary by our old sparring partner Morgan Matthews was very entertaining, but the bigfoot hunters that he chose to showcase came over as venal, stupid or mentally ill, and I seriously doubt whether there is anyone who actually thinks that this documentary added anything positive to the hunt for manbeasts in North America.

All film making is artifice, and I always thought that most people understood that, but apparently not.

On top of this, newspaper coverage of cryptozoological matters is often highly skewed, and I go into more detail about this in the excerpt from my Annual Report which follows.

Against such a cultural background it is not really surprising that fewer and fewer serious people become involved in a subject which is usually portrayed in the media as being nothing but a topic for facile entertainment.

We have talked a hell of a lot about how cryptozoology *is* portrayed in the media, but not yet discussed how we feel that it *should* be.

Heuvelmans himself said in one of the early issues of *Cryptozoology* (I can't find the exact citation today, because my elderly mother-in-law is currently using the CFZ library as her bedroom, and I would rather respect her space) that Cryptozoology was *not* the study of monsters, but the study of *unexpected* animals, and my personal view is that whilst the study of monsters is undoubtedly an interesting one, it does not come under the purview of Cryptozoology. I would also suggest that if the yeti - for example - *does* exist, it is - as Heuvelmans described - a shy and gentle great ape and not a monster at all.

Whereas, in these very pages, I do describe creatures as 'Sea and Lake Monsters', this is in reference to their monstrous size, not because of any attempt to imbue them with the more unpleasant of human characteristics.

So what is cryptozoology for?

Cryptozoology is the stuff of unexpected animals, and ethnoknown creatures presently unrecognised by science. Once upon a time every zoologist, and every naturalist who was interested in animals rather than plants, was a cryptozoologist, and I think that it would be a positive move if we were to go along that path again.

For some years I have had the suspicion that cryptozoology is *not* actually a science. Rather it is a description of an area of interest, rather like Natural History, and that a cryptozoologist, is more like a naturalist than a mainstream zoologist. I know that I am.

However, I am still interested in monsters.

I am still interested in vampires, werewolves and the ilk, and I cannot deny that I am interested in the more peculiar aspects of their kin, such as the Cornish owlman. After all, the CFZ could well be described as 'The House that Owlman Built' as my reputation was largely built on my 1997 book *The Owlman and others* which - to this day - remains my most popular book and has sold far more than everything else I have authored altogether.

This is the very reason why I founded the CFZ in the first place, and called it the Centre for *Fortean* Zoology, rather than mentioning cryptozoology anywhere in the title. However, it cannot be denied that cryptozoology is the major part of what we do, and that we are not the only organisation of our kind in the world, but one of the biggest, if not *the* biggest.

But what are we for?

That is the question which has been worrying me for the past year and a bit. I don't want to appear elitist at this point, but things were so much easier when cryptozoology was a minority interest. When it was the purview of amiable, though relatively grounded and intelligent eccentrics like Richard Freeman and myself, it was much easier to explain, and - for me at least - if you want to look at it in terms of social role valorisation, was a social role which I was happy to fill.

When it became devalued, and in my mind cheapened, it no longer was.

I am perfectly aware that this may seem snobbish to the people who are reading this, but in a cryptozoological community where people are continually sniping at each other, backbiting, and climbing all over each other in a desperate attempt to sell more books or get more people to visit their websites, and where an increasing number of people seem to believe that bigfoot employs a 'cloaking' mechanism similar to the Romulan cruisers on *Star Trek: The Next Generation*, neither the scientist, nor the anarchist hippy who just wants to live in peace love and harmony, or the anti-capitalist in me, is happy being part of that milieu.

So what is the CFZ for?

I believe that our role is exactly the same as it always was: to act as a clearing house for information, to promote excellence in research, and to provide an unbiased and neutral forum where researchers, pundits, and everyone else can meet and talk in safety.

There are other cryptozoological forums on the Internet, some of which are considerably bigger than we are, but some of them are veritable bear pits where only the thickest skinned dare venture without some hefty protection.

A couple of years ago I filled the 'father of the bride' role at Lizzy Clancy's wedding, and afterwards I followed one of my family failings and decided to preach a sermon (for those of you who didn't know my brother is a Lt Col in the British Army Chaplain's Corps). I took as my sacred text those immortal lines from *Bill and Ted's Most Excellent Adventure*: "Be excellent to each other, and *party on dudes*!"

And that is, in a way, another thing that the CFZ is for. I have been criticised over the years for calling our annual convention "The Weird Weekend", and not "The International Symposium for something or other". And

Raines Explains

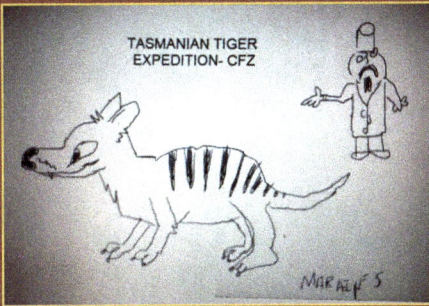

TASMANIAN TIGER
EXPEDITION- CFZ

that is for one important reason that should be tattooed on the foreheads of everyone who criticises this decision. THIS IS SUPPOSED TO BE FUN DUDE! Nobody makes any money out of this, no-one gets any more than the most transitory fame or glory out of it, so IF IT IS NOT FUN, IT IS NOT WORTH CONTINUING.

As far as the other half of Bill and Ted's most excellent axiom is concerned, another reason the CFZ does what it does is to BE EXCELLENT TO EACH OTHER. There are people involved with the organisation, me included, who have mental and physical health problems which - in the minds of many people - preclude them from having any worth as researchers, writers or contributors in any shape or form. This is patently stupid, and it is another reason for the CFZ's existence that we give a voice to people who otherwise would not be likely to have one.

We have a team of volunteer editors, copyproofers, and typesetters who work together with our authors to produce the best possible product. Sometimes this turns round and bites us on the bum. Last year an Asian author with whom we were trying to publish

a book, withdrew calling us racist and accusing us of insulting his family just because one of our editors wanted to make some perfectly justified minor changes to his manuscript.

My message to him, and to all the other people who have decided over the years that somehow CFZ Press is a sinister overlord trying to Bowdlerise their prose - self publishing software is freely available. If you think you can do it better, feel free.

I am also appalled by the behaviour of certain pundits within the community who have refused to help school kids when they write to them asking for information, or to do an interview for a project. Lighten up you idiots. In the grand scheme of things none of us is of any real importance; all we can try to do is to put more into the world than we take out, and that precludes behaving like an arrogant idiot when a child, or an adult with mental health problems comes to you for assistance.

And here, I find I have filled up six pages of an editorial that usually only takes up two, and that I haven't even mentioned our library project, the community arts and publishing programme, or our educational work with or without the Small School in Hartland.

But I tell you one thing: It feels good to be back!

Slainte,

Jon Downes (Director, CFZ)

THE CENTRE FOR FORTEAN ZOOLOGY
www.cfz.org.uk

A LEGAL MATTER

Relevant extracts from my 2014 Annual Report

You will probably have noticed that this past year the CFZ has done much less than usual, and that many of the activities that we have done for years have gone to the wall. There has been no issue of *Animals & Men*, no Yearbook, and only a handful of episodes of On The Track. Even the Weird Weekend was smaller than usual.

It would be easy to blame this on the recession. After all, despite the Government's talk of the 'Green Shoots of Recovery' they are hardly visible in what remains of our High Streets. Our Christmas sales this year, for example, have been respectable but hardly anything to write home about.

But it is not the recession that is to blame, at least not directly, but in many ways it is the effects of the recession on people who have hitherto given up their time for the CFZ. Some people who have hitherto had free time to do volunteer work with the CFZ now are to busy to do so, some people who because of their disability have previously been on State Benefits, have now been forced to take up demeaning and often health-threatening jobs which preclude them doing anything else, and across the board prices have risen - the most important from our point of view being postage, which is so prohibitively high now that it precludes many of the things that we have taken for granted during the twenty two years that the CFZ has been in existence.

There has also been an awful lot of deliberate malice about in the last few years. Looking back at the history of the CFZ I think that it was by far its most enjoyable back when Graham, Richard and I were on the breadline. People truly pulled together with us to make good things happen. However, it all changed once I inherited a relatively minor sum of money. As soon as my bank account was in the black, various people whom I had known for years decided that it was their bounden duty to divest me of it.

Before we go any further let me reassure you that the core team of Richard, Graham, Corinna and myself are still as thick as thieves, and I am sure that we shall always be so. However, the eagle-eyed amongst you will probably have noticed that at this year's Weird Weekend there were several old faces missing. Some, like Max Blake and Jess Heard had perfect reasons for not being there, and will, I am sure be back in the future. However, others were not there for far more sinister reasons. On the whole their presence wasn't missed and my enjoyment of the event was far higher for their absence, because these are people who have been running me ragged for years, and in several cases been robbing me blind. The machinations of one couple, for example, have this year cost me something between fifteen and twenty thousand pounds, which is more than my entire savings.

For much of the last twenty two years I have been propping up the CFZ from my own personal income. I am not complaining about this; I did it gladly, and will continue to do it gladly. But because of the aforementioned couple, whom I treated like friends and family, and who outmanoeuvred me at every turn, my personal income is now over five hundred quid a month lower than it was in the autumn of 2013, and I simply cannot afford to continue propping up the CFZ until my situation is improved.

This sounds like it is a precursor to me shutting up shop, but I assure you that it is nothing of the sort. The CFZ will continue much as it always has, and I have no intention of closing it down. But things have to, and are going to, change.

The cryptozoological community has changed massively in the past few years, and not necessarily for the better. For years cryptozoology has been an esoteric and rather arcane practice, in which only a small, intelligent and literate minority were interested. Now, because of the success of American TV programmes like Finding Bigfoot, it has become mass market entertainment and this is not necessarily a good thing. Not at all.

Now I am not being an intellectual snob (or any other kind of snob) here. I abhor that kind of elitism and self-aggrandisement, but the renewed interest in the subject in mystery animals that has come in the wake of the success of various mass market television programmes has done much to destroy a lot of the credibility that people like me have been working for years to build up.

Bryan Sykes et al wrote: *"Rather than persisting in the view that they have been 'rejected by science', advocates in the cryptozoology community have more work to do in order to produce convincing evidence for anomalous primates and now have the means to do so".* I agree with them entirely, and would also like to publicly reject any suggestion that cryptozoology has somehow

been 'rejected' by mainstream science, but it is certainly true that cryptozoology has been rejected, or at least shunned, by some scientists. Even scientists who are interested in the subject are discouraged in some cases from being involved. I shall not cite individual cases here, because this is neither the time nor the place, but believe me I could do. But in the current climate where cryptozoology is quite widely seen as nothing but the source material for yet another facile gameshow how can you blame them?

In these days when despite having been exposed as a hoaxer TWICE there are still some people who believe Rick Dyer's latest claims that he has shot a bigfoot, and where some people still believe the nonsense spouted by Melba Ketchum, the subject hasn't got a chance. However, it is not just the propagation of pernicious nonsense by people who should know better. There is an insane amount of negative reporting as well. *The Guardian*, for example, recently printed an article about the paper by Bryan Sykes et al claiming:

"DNA analysis indicates Bigfoot may be a big fake - New genetic analysis of 'yeti' hair samples reveals they actually originated from dogs, horses, bears or other known mammals".

Admittedly the Sykes paper did reveal that the samples originated from dogs, horses, bears or other known mammals, but nowhere did it even hint that this could be as a result of deliberate fraud, which is what a "Big Fake" would perforce have been. The newspapers just made that bit up!

Even the Loch Ness Monster wasn't immune. In November people from The Woodland Trust suggested that the most recent photographs and film purporting to be the monster could just be "logs floating on the water." Within a few days the "Nessie was a fake all along" headlines were appearing the very same newspapers which a few days before had been touting the very same unlikely videos as the "definitive proof" of the monster.

This continual tirade of negativity and nonsense has discouraged many people from being involved in the serious end of cryptozoology. It irritates and even angers me, but I think in these decadent times we need the CFZ to be active more than ever.

The cryptozoological community is also changing, and in several ways is not as nice a place to be as it was a few years ago. The advent of ubiquitous social media means that everyone can and does state their opinion about various things several times a day. I know that I do, and I applaud the fact that other people can and do likewise.

However, sometimes this new freedom can and does turn round and bite us on the bum.

Most weeks I get a complaint from some earnest and well-meaning CFZ member and/or sympathiser, complaining about the public behaviour of one or more senior member of the CFZ. This usually means that they have said something on Facebook about which the other person disapproves. These subjects include, but are not restricted to, animal rights, hunting, sexism, homophobia, abortion, religion or some other related subject. I even receive complaints that this member or that member has posted a picture of a scantily clad Hollywood starlet on their homepage, and I have to spend a lot of my time trying to think up polite ways of telling the complainant that not only do I not care, but I couldn't do anything about it even if I wanted to.

The CFZ is a loose confederation of people broadly interested in one or more aspects of mystery animals, Forteana, and the culture surrounding them. This doesn't mean that we all think, believe, or like the same things. I do not choose to put pictures of page three girls on my Facebook page, because to my mind it is vulgar to do so. I am also 55 and happily married, but I have no intention of criticising my friends and colleagues, let alone people I hardly know, because they choose to do so. It is simply none of my business.

A whole generation of like-minded folk quite rightly think that the advent of the Internet and its accompanying technology heralded the manifestation of the concept of the Global Village which was promulgated half a century ago by people such as Marshall McLuhan. However, although I live in a village, and would much rather do so than live in a city like I did for twenty years, I am not blind to the fact that villages have their disadvantages, as do global villages.

The internecine squabbling over relatively minor things such as I describe above has driven many people away from the cryptozoological community, and I will be the first to admit that mine and Richard's high profile protesting about subjects like the badger cull, and the proposed re-legalisation of fox hunting, has annoyed and alienated some people. But unlike someone making sexist jokes on their Facebook pages, these are subjects about which I am prepared to nail my colours to the mast. And as the CFZ is an animal welfare organisation, I feel that I have no option but to do so.

I have also been told that the CFZ has lost support in some quarters because I am a vegetarian, drink to excess occasionally and smoke cigarettes. How can I expect anyone to take me seriously when I do things like that? Personally, I don't think it is anyone's business except for mine and my wife. So there.

Another problem with village life is the way that various cliques squabble with each other in constant games of one-upmanship. I have known this happen in the village in which I live, but have never known it as prevalent as it is at the moment within the cryptozoological community. The CFZ has been around for over twenty years and is undoubtedly one of the largest cryptozoological organisations that has ever existed, and I like to think that we have been largely a success. However, the cryptozoological community is full of people who would rather be divisive and petty in an attempt to sell a few more books or get a few more visitors to their websites. The false sense of intimacy that social media engenders is conducive to such idiotic squabbling and divisiveness. It is a game which I refuse to play, but I am afraid to say that others do.

The final social change which is proving very harmful to the traditional model of the CFZ is the fact that more and more people expect to get stuff for free. This has already irrevocably changed the music industry, and I am the first to admit that back in 2001 when I first logged on to Napster I did my bit to promote this sad decline. But the traditional business model in publishing, especially where magazines are concerned has now changed forever.

I am a stubborn old so and so, and have always tried to stick to my guns. I don't like e-books, and I don't like the way that electronic publication, especially through websites, has largely replaced hard copy publishing in some quarters. The CFZ has always insisted that its flagship publication *Animals & Men* would never become a digital-only magazine. So it won't, but it is going to change.

As you may know I have been editor of the Gonzo Multimedia weekly music magazine, the *Gonzo Weekly*, for over two years now, and I was also editor of *UFO Matrix* during its final four issues. During that time I have learned a lot more than I thought that I ever would about digital publishing.

As of this issue *Animals & Men*, will be produced as an e-magazine and available for free online. However, unlike *Gonzo Weekly* it will also be published simultaneously (or within a few days either way) as a hard copy magazine available from us or via Amazon and other online outlets. We intend to start this new model as quickly as possible but it will, perforce, mean a very drastic change in the way that CFZ membership works.

For the first time since 1994 there will be no subscription model for *Animals & Men*. The membership package will now consist of a monthly newsletter, a book, and a series of discounts on CFZ publications and tickets to the Weird Weekend. Details of is new package will be announced in the next few days.

All current subscriptions will be migrated to the new format, and a credit note issued for any outstanding monies held by us. That credit note can be redeemed against any of our publications including *Animals & Men*, so although we are changing the model, subscribers can still stick to the current plan if they wish.

More details will be announced in the next few days. Whilst it will be completely free to read or download the magazine in a pdf flipbook format, there will be an opportunity to sponsor the magazine via Patreon.

Initially the important thing is to return to the quarterly format which we have not achieved for many years, but in the medium term the plan is to make *Animals & Men* a monthly publication. But baby steps first.

Newsfile

New & Rediscovered

rylandsi

male

female

Saki Malarkey

Sakis, or saki monkeys, are any of several New World monkeys of the genus Pithecia. They are closely related to the bearded sakis of genus Chiropotes. They are small-sized monkeys with long, bushy tails. Their furry, rough skin is black, grey or reddish-brown in colour depending upon the species. The faces of some species are naked, but their head is hooded with fur. Their bodies are adapted to life in the trees, with strong hind legs allowing them to make far jumps. Sakis reach a length of 30 to 50 cm, with a tail just as long, and weigh up to 2 kg.

Five newly discovered species from Brazil bring the total number of species to sixteen. "I began to suspect there might be more species of saki monkeys when I was doing field research in Ecuador," said lead author Dr Laura K. Marsh, primate ecologist and director of the Global Conservation Institute. "The more I saw, the more I realised that scientists had been confused in their evaluation of the diversity of sakis for over two centuries."

The five new species are found in Brazil, Peru, and Bolivia – three of them are endemic to Brazil and one to Peru. This revision increases the number of primates in Brazil to 145; the highest diversity for any single nation.

SOURCE
http://www.wildlifeextra.com/go/news/Saki-monkey-discovery.html#cr

'Chilla Thriller

Chinchilla rats or chinchillones are members of the family Abrocomidae. This family has few members compared to most rodent families, with only 9 known living species. They resemble chinchillas in appearance, with a similar soft fur and silvery-grey colour, but have a body structure more like a short-tailed rat. They are social, tunnel-dwelling animals, and live in the Andes Mountains of South America. They are probably herbivorous, although this is not clear.

The Machu Picchu arboreal chinchilla rat (Cuscomys oblativa) was first described from two enigmatic skulls discovered in Incan pottery sculpted 400 years ago.

Dug up by Hiram Bingham in 1912, the skulls were believed to belong to a species that went extinct even before Francisco Pizarro showed up in Peru with his motley army. Then in 2009, park ranger Roberto Quispe found what was believed to be a living Machu Picchu arboreal chinchilla rat near the original archaeological site.

"In conservation biology, this type of rediscovery is called the Lazarus effect," said a team of Mexican and Peruvian scientists, who have sought to confirm Quispe's discovery.

The scientists - led by Horacio Zeballos the curator of the department of mammalogy at the Museum de Arequipa and Gerardo Ceballos from the Instituto de Ecología of the Universidad Nacional Autónoma de México - headed into the field in 2012 and tracked down the elusive mammal in the cloud forests near Wiñayhuayna, another archaeological site on the Inca Trail heading towards Machu Picchu.

SOURCE
http://www.theguardian.com/environment/2014/sep/26/extinct-cat-sized-chinchilla-found-in-shadows-of-machu-picchu

Please look after this bear

One of the most encouraging zoogeographical events of the century so far is the resurgence of large mammals in Europe.

As reported in our last issue wolves are back in Denmark for the first time in 200 years, and our Danish correspondent Lars Thomas tells us that there have been records of lynx very near to the border with Germany. In the east of the continent there is some spectacular news. Scientists have captured what is believed to be the first photographic evidence of brown bears within the Chernobyl Exclusion Zone (CEZ).

S.GASHCHAK/CHORNOBYL CENTRE

Camera traps, used by a project assessing radioactive exposure impacts on wildlife, recorded the images. Brown bears had not been seen in the area for more than a century, although there had been signs of their presence. The exclusion zone was set up after an explosion at the Chernobyl nuclear power plant in Ukraine in April 1986.

"Our Ukrainian colleague, Sergey Gashchak, had several of his camera traps running in one of our central areas over the past few months in order to start to get a feel for what (wildlife) was there," explained project leader Mike Wood from the University of Salford.

He told BBC News that data retrieved from one of the cameras in October contained images of a brown bear. "There have been suggestions that they have existed there previously but, as far as we know, no-one has got photographic evidence of one being present on the Ukrainian side of the exclusion zone," Dr Wood said.

SOURCE
http://www.bbc.co.uk/news/science-environment-30197341

The Frog Chorus

Frogs are generally in the news these days because of the appalling predations of the

chytrid fungus. But new species of frog are being discovered across the globe. Just a few examples:
Scientists say they have discovered 14 new species of so-called "dancing frogs" in southern Indian forests.

Indian biologists say they found the tiny acrobatic amphibians, which earned their name with the unusual kicks they use to attract mates, declining dramatically in number during the 12 years in which they chronicled the species through morphological descriptions and molecular DNA markers. They breed after the yearly

monsoon in fast-rushing streams, but their habitat appears to be becoming increasingly dry.
"It's like a Hollywood movie, both joyful and sad. On the one hand, we have brought these beautiful frogs into public knowledge. But about 80% are outside protected areas, and in some places, it was as if nature itself was crying," said the project's lead scientist, University of Delhi professor Sathyabhama Das Biju. Biju said that, as researchers tracked frog populations, forest soils lost moisture and perennial streams ran inexplicably dry. He acknowledged his team's observations about forest conditions were only anecdotal; the scientists did not have time or resources to collect data demonstrating the declining habitat trends they believed they were witnessing.

SOURCE
http://www.theguardian.com/environment/2014/may/08/dancing-frog-species-discovered-in-indian-jungle-mountains

Most frogs lay eggs and although some species give birth to froglets, newborn tadpoles are new to science. This species of fanged frog lives on Sulawesi Island in Indonesia, and zoologists had chased it for decades because they suspected it would show this unique behaviour. An international team has now described it for the first time, in a study published in the journal *Plos One*.

Dr Jim McGuire from the University of

California, Berkeley, actually thought he was holding a male frog the first time he witnessed a birth. In fact, he had in his hands a pregnant female and, suddenly, a clutch of brand new tadpoles. Nearly all the world's 6,000 frog species use external fertilisation: the female lays eggs during mating, while the male releases sperm to fertilise them. "But there are lots of weird modifications to this standard mode of mating," Dr McGuire said. "This new frog is one of only 10 or 12 species that has evolved internal fertilization, and of those, it is the only one that gives birth to tadpoles, as opposed to froglets or laying fertilized eggs."

SOURCE
http://www.bbc.co.uk/news/science-environment-30643756

Scientists have confirmed that a frog found living in New York City wetlands is a new species. Jeremy Feinberg, of Rutgers University in New Jersey, who led the study, first reported the discovery when he heard their "very odd" chorusing call. Teaming up with genetics experts to confirm the finding, Mr Feinberg has now published the discovery in the journal *Plos One*. The researchers realised the frog was a new species when they heard its call. It is the first new frog species found in the region for nearly 30 years.

Mr Feinberg told BBC News he knew he might be on to something when he heard a group of them calling in chorus at a wetland study site on Staten Island. "Frogs have very stereotyped calls within a species, so I knew this was different," the ecologist told BBC News. "But it took me two years to find someone to partner with me on the genetics side." He believes the frog, named *Rana kauffeldi* - a leopard frog - probably once inhabited Manhattan, so it had been seen before. But it was assumed to belong to a similar-looking, previously known species of leopard frog (so named because of its spots) found in the same area.

SOURCE:
http://www.bbc.co.uk/news/29792309

A new species of tree frog has been found in Madagascar and immediately rated as "highly threatened". University of Bristol student Samuel Penny was studying on the island when he found the species, which has since been named 'boophis ankarafensis'.

Some 56 individuals were detected near hidden streams in the Ankarafa Forest. Scientists now say the species should be classified as critically endangered due to a "continuing decline in the quality and extent of its habitat".

SOURCE
http://www.bbc.co.uk/news/uk-england-bristol-28911783

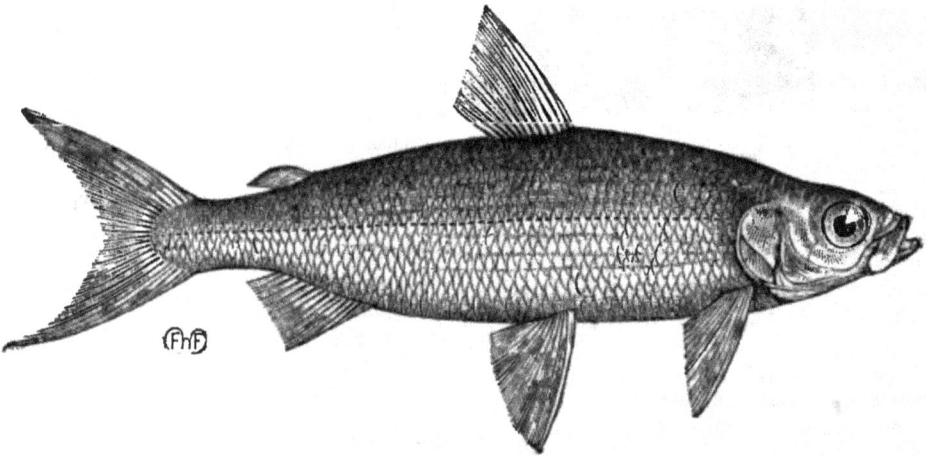

Vendace (*Coregonus vandesius*). (From Loch Maben.)

A Fishy Story

Two adult vendace, Britain's rarest freshwater fish and a relic of the last ice age, were found in Bassenthwaite Lake in north-west England last month, more than a decade after being declared 'locally extinct'. Last year, a single young vendace was recorded during the annual fish survey. Dr Ian Winfield from the Centre for Ecology & Hydrology leads the annual fish surveys of Bassenthwaite Lake. He said, "Finding adult vendace in 2014 following the recording of a young fish in 2013 is excellent news giving great encouragement to everyone involved in the restoration of Bassenthwaite Lake and its fantastic wildlife."

The fish community of Bassenthwaite Lake has been monitored by the Centre for Ecology & Hydrology since 1995 in a collaborative project previously with the Environment Agency and now with United Utilities. Each year, the abundance and composition of the fish community is assessed using state-of-the-art hydroacoustics (echo sounding) combined with limited netting. Dr Winfield added, "The news for Bassenthwaite Lake is about as big as

it gets for rare fish. I am certain that other adults remain in the lake. I also think that such fish will spawn this winter, but I'm unsure of how egg incubation will go given persistent sediment problems at the lake."

The 2014 fish community survey of Bassenthwaite Lake was carried out on 10 September 2014 and recorded two post-underyearling vendace which would probably be old and large enough to spawn as adults during the coming spawning season of late 2014. There are a number of possible origins for the vendace found in Bassenthwaite Lake in 2013 and 2014. One theory is that vendace have actually survived in Bassenthwaite Lake below the limit of detection for the last decade and may now be increasing in abundance. A second is that fish may recently have arrived in Bassenthwaite Lake by moving down the River Derwent from the population in Derwent Water. Recent DNA analysis suggests that the underyearling fish recorded in 2013 originated from Derwent Water, but similar tests have yet to be performed on this year's adults.

SOURCE
http://www.ceh.ac.uk/news/press/adult-vendace-bassenthwaite-lake-september-2014.asp

Man Beasts (BHM)

Is 'Rick Dyer' Cockney rhyming slang?

Twice now, Rick Dyer has managed to fool a proportion of the bigfoot research community into believing that he was in possession of the body of one of the fabled North American manbeasts.

On August 12, 2008, Matthew Whitton and Dyer released a press release and went on Steve Kulls' radio show Squatch Detective to announce they had a dead Bigfoot body in their possession. After initially leaking grainy footage that showed Bigfoot, they presented the carcass encased in a block of ice at a conference that was only open to the press. The two announced that they found the 7-foot-7-inch, 500-lbs creature while hiking in the north Georgia mountains in June. They also stated that they had

spotted about three other similar creatures after making the discovery. Tom Biscardi joined Whitton and Dyer for the news conference, stating "Last weekend, I touched it, I measured its feet, I felt its intestines" and lauded its authenticity. The cadaver turned out to be a Hallowe'en costume stuffed with roadkill - possums according to several reports

Biscardi has already been involved in what appears to have been a hoax three years prior to this, although it is not clear whether he was a knowing participant in the Dyer/Whitton hoax. Both Biscardi and Dyer appeared in Morgan Matthews' 2013 film *Shooting Bigfoot*. I know Morgan and have had business dealings with him, so I feel it is not appropriate for me to comment on his film. However, Red Barracuda' describes the film on the IMDB:

"Morgan Matthews follows three separate groups of Bigfoot hunters on individual treks. The results are edited together. Firstly there is Tom Biscardi, a man

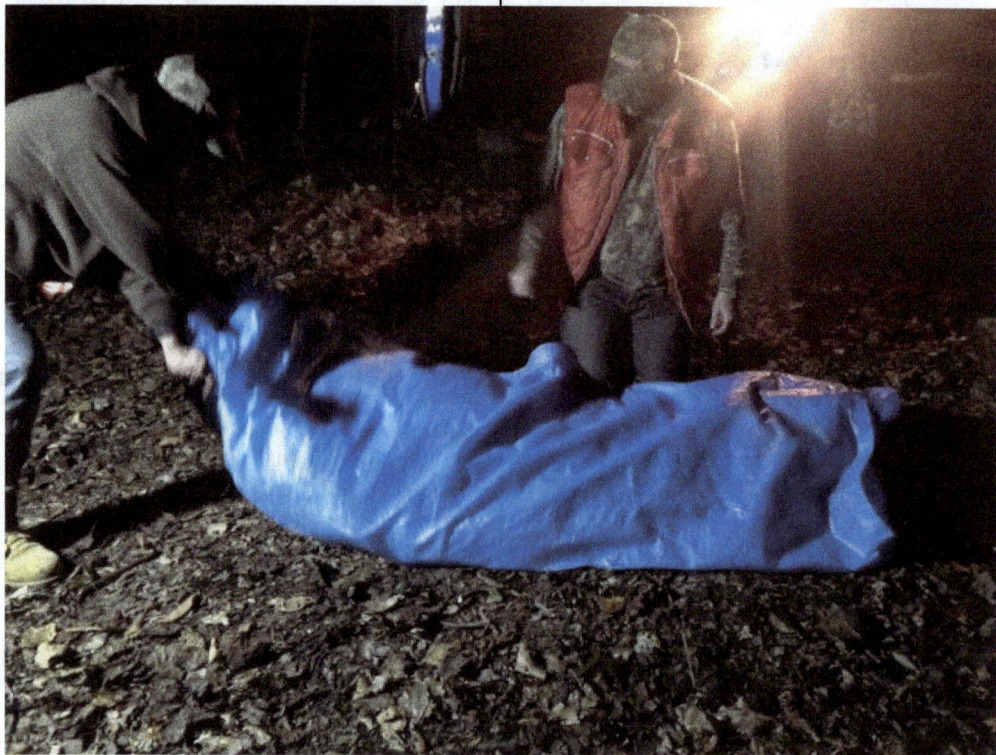

who has spent decades making documentaries about the subject. He is quite a highly strung fellow and somewhat self-important. Many of the funniest moments revolve around him and he has some very amusing dialogue throughout. Secondly there is Dallas and Wayne who are a couple of elderly hillbillies. They are much more sympathetic characters than Biscardi and it's hard not to feel a little sorry for them in their dead-end obsession. They drive out into the woods to play cassettes and make animal noises in an attempt to lure in the beast. The walls of their home are full of photographs of evidence but truthfully they are photographs of nothing. Their obsession is given a little context in that a work injury left Dallas unable to continue in his job leaving him trying to find meaning for is life in the hunt for Bigfoot; it's quite sad, although he does seem a undeniably little mentally unstable especially when he claims he has an affinity with Bigfoot because he has animal DNA as a result of a sheep bone being implanted in his head in order to seal a wound. Lastly, there is Rick Dyer who is a self-styled 'Bigfoot Tracker'. He

really seems to be a somewhat dangerous man and appears to engage in games with Matthews to try and freak him out in the middle of the night. The fact that he spends most of his time carrying a loaded rifle doesn't exactly help matters. To make the situation worse they encounter a young homeless man in the woods who may or may not be secretly in cahoots with Dyer. But even if not, this guy gives off the impression of someone to keep well away from. One night he pitches up at the camp site with his dog horribly gashed at the neck. His ambivalence on the matter made you wonder if he was the one who actually did it. Anyway, this strand of the film is the only part that actually ends with any conclusion. Surprisingly, it's a pretty scary one. Although it does push the documentary onto the 'is-it-a-mockumentary?' side of the fence."

Following the events of that movie Dyer claimed to have killed a Bigfoot-like creature in San Antonio, Texas, in early September 2012. According to Dyer, he lured the Bigfoot out using "pork ribs from Wal-Mart", doused in a special barbecue sauce, that he attached to trees. Of the experience, he told *Esquire* magazine "We nailed 'em all around the trees, and then that night we heard Bigfoot come back. I chased him down in the middle of the night. I shot him once, he ran, I shot him again."

He declared than an unnamed university in Washington state had tested the creature's (which by this time was nicknamed 'Hank') DNA and told Dyer that it was an unknown species. Dyer's accomplice and self-proclaimed bigfoot skeptic, Allen Issleb ("Musky Allen") of Wauconda, Illinois, claimed to have inspected Dyer's bigfoot in Las Vegas in February 2013 and proclaimed it to be the

announced at a press-conference in late February or early March.

This is what we know so far:

- A Medical Doctor has verified the authenticity.
- Walter Shrum (Bigfoot Song Writer and Musician) drove 4 hours to see the body. (He also verified the authenticity).
- A total of 9 people participated in the expedition.
- The bodies are in Georgia being prepared for the press-conference next year.

Keep it here for the latest on this saga!"

However, at the time of going to press, the Project Sasquatch website seems to have been taken down, and no more has been heard from Dyer or anyone else about this new claim. Funny that!

real thing. This claim resulted in many people getting sucked into Dyer's scam. Dyer called the creature Hank and started touring the body around the United States, charging people to view it. To view the body, which lay beneath Plexiglas in a wooden coffin, adults were charged $10 and children were charged $5. By Dyer's own admission the tour eventually pulled in close to $60,000.

After 'Hank' had been shown to be a fake, Dyer said on his Facebook page: "From this moment on, I will speak the truth! No more lies, tall tales or wild goose chases to mess with the haters! I never treated anyone bad, I'm a joker, I play around, that's just me."

Less than eight months later, projectsasquatch.com announced:

"On October 25, 2014 Bill Burk , Rick Dyer and Tim Fricke killed a Bigfoot in northern Pennsylvania. Details of the event will be

Mystery Cats

Jekyll and Hide?

Jekyll Island off the coast of Georgia, USA was first 'discovered' by the Spanish in 1510, and the notorious Juan Ponce de Leon who is one of the subjects of my 2007 book *The Island of Paradise* was an early governor.

Until now, bobcats (*Felis rufa*) have only been rumoured to exist on the island, with the only concrete evidence of their existence being a photograph from about a century ago showing bobcat pelts nailed to the wall of a gamekeeper's hut.

Now, after an absence of a century there is finally concrete evidence for their existence; a recent survey by the DFW, trying to monitor numbers of white tailed deer, has shocked island experts by revealing two separate photographs of a bobcat from different locations on the island.

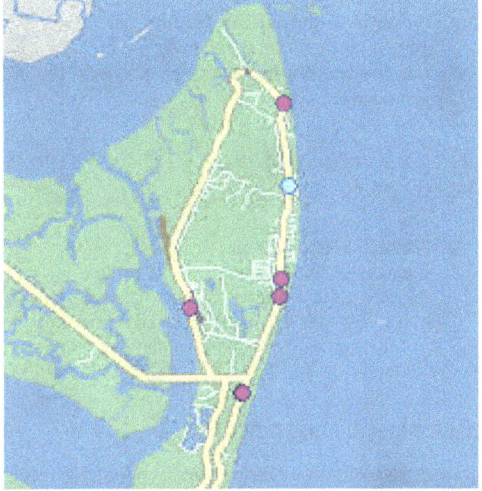

This is where it gets interesting, because although on the surface it might appear that this is a cut and dried case of a mystery animal turning up where experts said there weren't any, it is quite possibly not that at all. Conservation Director Ben Carswell is quoted as saying:

"It's the first definite, confirmed documentation of a bobcat on the island ever. We have no way to be sure whether

this animal showed up recently on Jekyll. They're such secretive animals; it could be this one and others have been out here for some time"

The *Mysterious Universe* website quite rightly points out that:

"Bobcats were believed to have lived on the island until being wiped out by hunters in the early 20th century. This one most likely swam from the Georgia mainland across a march or possibly ran there via the five-mile causeway."

But it is still good news.

Danish Predator

In early December Lars Thomas wrote:

"Danish media are in a state of mild panic today following the find of a very dead roe deer in Dyrehaven, a very popular Danish deerpark/woodland - in fact the most visited forest in Denmark.

It is located just north of Copenhagen, and is used by hundreds, if not thousands of people on a daily basis - especially in summer. It is also home to a large deer-population that are strictly controlled by the authorities. You can find red deer, fallow deer, sika deer and roe deer in quantities. Although today it is in quantities -1, as a big roe deer calf was found killed.

As the picture shows, someone or something had basically ripped the deer's head off. The authorities are at the moment working from the assumption that the culprit is a large dog, although the actual kill does not confirm with a standard dog kill, nor for that matter with a wolf kill. DNA-samples has been taken, but the results will probably not be in until after the weekend. Until further notice people are being advised to be careful should they meet a big stray dog in the area. The Danish anti-wolf lobby is of course already up in arms, wanting every Danish wolf shot, just in case it is one of them - never mind the fact that the closest sighting of a wolf is some 300 km west of Copenhagen.

Richard Freeman immediately claimed that it was the work of a troll, but the next day Lars wrote an update:

"No wolves, no trolls, only dogs!!!

As my last blogpost states, a dead headless young deer was found on Friday in a very popular deerpark/forest just north of Copenhagen. It created quite a stir, as it wasn't just killed. It's entire head and neck had been ripped off. Since wolves were officially added to the Danish fauna in 2012, everybody has been a little jittery, so of course the newspapers went berserk with stories about whether it could actually be a wolf. And the conspiracy theorists started coming out of the woodwork suggesting that a wolf had been deliberately released (it is after all a deerpark which is completely fenced in) or that somebody had brought the dead deer from somewhere else to get things going a little. And by the way, it was not as I stated in my earlier post a young roe deer, but a young fallow deer.

Anyhow - DNA-samples were collected, and the results are now in. The attacker was a dog, not a wolf. This of course has in no way diminished the attack from the anti-wolf lobby, who for some reason thinks a small handful of wolves are far more dangerous than any number of dogs. I am certain we haven't heard the last of this."

Aquatic Monsters

Down to the lake I fear

Some years ago an *Animals & Men* reader emailed me suggesting that Google Earth would be a useful medium to use when searching for cryptids. He pointed out that there are certain places where one can see stranded whales, and their living counterparts out to sea.

Since then there have been a number of stories in the world's newspapers claiming that 'monsters' have been spotted on Google Earth images of Loch Ness, and on one occasion Loch Lomond. These are all fairly

obviously pictures of boats with their attendant wakes. There has also been a claim that a crab of truly mammoth proportions can be seen next to a pier on the sea front of the Kent town of Whitstable, but that is another story (and one which, furthermore,

should be taken *cum grano salis*) entirely.

However, now an image that seems far more interesting *has* turned up on Google Earth images of Loch Ness.

The *Daily Mail* online says:

> "A Swedish monster-spotter has won a £2,000 prize for a picture he took of the legendary Scottish creature - without leaving his home in Stockholm.
>
> Bjarne Sjöstrand, 52, spotted a long, thin object in the water on Google Earth, after he logged on to his computer 800 miles away.
>
> Mr Sjöstrand, who has never visited Scotland, won the best Loch Ness Monster sighting of 2014, in an annual competition run by bookmaker William Hill. The systems administrator was at home in Stockholm when he spotted a long thin object in the water on a Google Earth image taken above Horse Shoe Scree on the loch's southern side.
>
> Mr Sjöstrand said: 'I am very interested in anything regarding Nessie and the history of Loch Ness.

> 'The reason I found this image on Google Earth was that I was sitting one night at home reading about Nessie and Loch Ness on the internet and thought I would check to see if I could see something from above - and that proved to turn out well.
>
> 'I have never been to Scotland but my hopes are that one day I will visit Loch Ness.'

Although I would like to stress that there are no CFZ Thought Police; we are a very broad church, and apart from the proviso that research is *never* done for political or religious ends, the organisation has people who believe all sorts of things, but the core members of the CFZ have always thought that it is most likely that a large proportion of the lake monster reports which have a basis in zoological fact, are of enormous eels.

This latest image would seem to be corroborating evidence of that theory. It is by no means conclusive, and may well turn out to be a drainage pipe or something else entirely, but - for once - an image has come out of Loch Ness which is roughly analogous to what we would have thought *should* be there, other than a picture which is either obviously a fake, or a mistaken identity, or something that if it is real has come out of left field entirely.

It will be interesting to see what happens next!

SOURCE:
http://www.dailymail.co.uk/news/article-2907932/Swedish-monster-spotter-finds-shape-waters-Loch-Ness.html

Down to the lake I fear (2)

SOURCE:
http://nerdalicious.com.au/ridiculam-mundi/
australian-adventurers-capture-loch-ness-monster/

And there's another one. 2014 was such a poor year for Loch Ness Monster sightings that the popular press had a field day claiming that Nessie (I *do* wish that people would stop using this ridiculous nickname, and claiming that the creature is female) has died.

Now it is only half way through January and there have been two new pictures already.

This one, to my mind, however, reeks of fish, not because I am claiming that the LNM is piscivorous, although it probably is, but because I think that this is a blurry photograph of a grey heron.

The story that accompanies the picture is from an Australian website which contains a mixture of factual and humorous stories, and it is not clear in which category this story falls. Sadly, I suspect the former.

Carl Marshall's Column

Introduction

Back in October 2012, while driving along the A46 near Evesham in Worcestershire, Andrew Badland, (below) passed a strange animal and caught it briefly in his headlights, highlighting it long enough to see the hind region clearly as it slunk along a hedgerow through into a neighbouring field out of sight. Andy described the animal as muscular and sandy coloured, just over two feet high and about five to six feet long, with a pronounced crook in its long tail. I interviewed Andy about his alleged encounter and the credibility of his account became more apparent as the interview progressed.

Transcribed Interview

Q. Can you describe to me, in as much detail as possible, what you saw when you were driving home that night?

A. I was driving late at night along the A46 near Evesham and I passed what must have been a big cat. It was a brownish sandy colour and had a very long tail that went down toward the ground and then turned up, sort of floating above it; the tail was the same colour. I see foxes almost nightly and this was definitely not one, the tail was too long and slim. I am also a hundred percent it was not a cinnamon badger [Sometimes erythristic badgers are locally called cinnamon badgers]. This was big and looked predatory and it definitely had a hook in the tail.

Q. That's interesting, this area is known for big cat sightings but ones usually concerning a melanistic animal.

Investigation of an alleged sighting of a puma-like animal on the outskirts of Evesham, Worcestershire.

A. No this was definitely a light, brownish kind of colour.

Q. Where exactly did this sighting take place and what date time was it?

A. It was last October; I was travelling down the A46 Sedgeborough bypass heading towards Tewkesbury, it must have been 10:45 [pm].

Q. Was there anything recognisable near the animal to give an indication of its size proportions?

A. Yes it was walking along the hedgerow on the roadside, it must have just crossed the road. If only I was two seconds further along the road, I would have seen the entire animal clearly.

Q. How large would you say it was?

[Andy crouched down on the floor and gestured a size, it would have been just over sixty centimetres to the animal's haunches.]

A. I couldn't tell you how long it was for certain because I missed its front end but it was probably about five to six feet.

Q. You mentioned the creature's tail - could you describe this feature in as much detail as you can remember?

A. Yes the tail looked long, probably about two, to two and a half feet long, sandy coloured with a definite hook in it, nothing at all like a fox or a dog. It looked to be almost as long as the cat's body and not overly thick.

[later I pointed out the crook in a German Shepherd's tail to which Andrew positively stated was much shorter, more bushy and

looked quite different to what he witnessed that night.]

Q. Roughly what speed were you travelling at the time of the sighting?

A. About forty [mph]

Q. How long did your sighting last in total?

A. About five seconds.

Q. I have to ask this but had you consumed any alcohol or had taken any other intoxicating substance at the time of the sighting that could have altered your perception?

A. No I was completely sober, I don't really drink and I wasn't even that tired!

Q. Can you remember seeing any distinct markings on the animal such as spots or stripes?

A. No, while I had it in the headlights it appeared to be all one colour!

Q. Did you get the impression that the animal you witnessed was in good condition or did it look unhealthy?

A. No it looked muscular what I saw of it, but it all happened quite fast. I saw its back end fairly clearly for a while and I'm sure about the shape and size of its haunches and tail. Also I thought just after the sighting that its hind feet seemed large to me, larger in comparison to say a dog or fox, but I didn't register this until after I had passed the animal.

Puma photographed in Belize by the author

Q. Have you heard of any other similar sightings being reported in this area?

A. Only the black cat that is sometimes seen around here.

Conclusions

Andy's encounter is interesting for two reasons. Firstly and most notably this particular area is already known for an anomalous big cat, but of the melanistic colour variation. So the fact that he reported a 'sandy coloured' animal with no distinct markings is unusual and of particular interest. Secondly Andy is a capable naturalist and knows both British and exotic wildlife well, having a keen interest in native species and also working with exotic animals for a number of years. He is adamant he would not mistake a domestic dog, fox, or any other British animal for a big cat, and when I made a point of showing him a dog breed that displayed a pronounced crook in the tail (German Shepherd) he pointed out that the animal he witnessed that night had a much longer and more conspicuous tail and also appeared to display what Andy later perceived to be disproportionate hind feet. Andy also later stated to me that the animal he witnessed was too large to be a domestic cat, even a very large breed like a Maine Coon. Some critics might automatically assume that Andy had seen a large (very large!) dog fox! Especially as he didn't see the animal's head, but Andy is an observant naturalist and I personally do not believe he would easily mistake a fox for a puma. The fact he described the tail as: "long, probably about two, to two and a half feet long... with a definite hook in it, nothing at all like a fox or a dog. It looked to be almost as long as the cat's body and not overly thick" and "had a very long tail that went down towards the ground and then turned up, sort of floating above it... the tail was too long and

slim" - doesn't sound much like the average fox in winter months!

So if what Andy saw was a puma is it likely that two species of big cats are sharing the same territorial range in Britain?

Well maybe, but to investigate this further we need to propose identities for these animals purely to make comparative predictions. Even though we should rightly be cautious when attaching specific identities to a series of sightings, and it is indeed true we can never be certain as to the identity of an unknown animal until a verified specimen is found, biologically studied and identified, in certain circumstances it is beneficial to theorise if only to make ecological predictions.

Worcestershire's Black Mystery Animal - An alleged large black cat has been reported in Worcestershire and surrounding border counties for a number of years so therefore there may be more than one animal responsible, possibly even successive offspring, as sightings of females with cubs have been reported, or more likely a series of wandering escapees. In general it is a large black animal that appears to be a Felid. It appears to be nocturnal but has also been witnessed diurnally, and occasionally crepuscular activity has also been witnessed and reported. The animal genuinely appears to be melanistic, so therefore most likely either a jaguar or a leopard with the latter being more likely as these are more suited to surviving in Britain's colder climatic conditions. Also melanistic leopards were one of the more popular exotic cats to privately own prior to the introduction of the Dangerous Wild Animals Act 1976. The most likely identity for Worcestershire's mystery black cat reports is a melanistic leopard of unknown sub-species. Leopard - *Panthera pardus spp.*

New unidentified Animal - This animal, reported by Andrew Badland, has only been documented once in this particular area and was recorded as large - approx two feet to the haunches, approx 5 to 6 feet in total length and sandy coloured. The identity I am proposing for this animal is a puma of unknown sub-species and provenance. Puma - *Puma concolor spp.*

If the animal Andy reported was indeed a puma then what can we expect next? Will there be further sightings? To answer these questions we first need to pose another. What would happen if a puma entered the established territory of a leopard and came into direct contact with it? This is a difficult question to definitively predict as these two species do not naturally come into contact anywhere in the wild - *P. concolor* being a new world predator and *P. pardus* from the old world (Africa and tropical Asia). Pumas are adaptable, generalist species that are found in most American habitat types and are known to have exceptionally large territorial ranges. In the Americas its range is from the Canadian Yukon down to the southern Andes of South America, making it the greatest of any large terrestrial mammal in the western hemisphere. They are a highly territorial species and known to survive in low population densities. See Florida Panther - *P. c. coryi*. Pumas are without doubt dangerous predators, although not always the apex predator in a particular habitat often yielding to the grey wolf *Canis lupus*, American black bear *Ursus americanus* and grizzly bear *Ursus arctos ssp*. and most notably here the jaguar *P. onca* - a species which is similar in size and territorial nature to a leopard (leopards are actually slightly smaller). So from this we could deduce that even though an escaped puma would quite possibly enter the range of an out of place leopard (which Andy's

encounter suggests) it would be extremely unlikely to come into direct contact with it, as a puma would be likely to avoid the latter either by living cryptically in the area for a short time while its vacant, or permanently leaving the territory in search of another. Maybe our mystery Felid has started a new territory that borders, and even overlaps, the alleged leopards.

Taking into consideration Andy's trustworthy nature and professional credibility it's possible he saw what he claims, so there may be more reports of puma-like animals in and around Worcestershire if the animal remains in the area. However if there are no more reports it can also be interpreted that the animal sighted above may have moved into another territory to avoid possible confrontation. Considering the relatively low numbers of big black cat sightings in my home county of Warwickshire maybe our alleged puma spends most of its time there. However, due to the lack of any credible data this is simply speculation.

My investigation into whether the West Midlands has a new big cat continues!

A special thank you to Andrew Badland for the above report, Maureen Ashfield B.A.Ed (Hons) for critiquing this article and to Richard Lamb C.Biol MSB for his invaluable zoological advice.

Artwork by Emma Cole

Carl Marshall works at Stratford Butterfly Farm and is a fine field naturalist. Over the past couple of years he has become a very enthusiastic member of the CFZ, and his quasi-Fortean view of British natural history fits in perfectly with my own. He was, therefore, the perfect choice as a columnist for the brave new *Animals & Men*, and we are proud to have him aboard.

Woodlice - many people are actually afraid of them. So much so that one lady refuses to have a bath or shower for fear that a woodlouse will drop on her head. This is pretty extreme, but we should take a look at the little fellow to satisfy ourselves that they are not our nemesis and even if they do land on us they will do us no harm.

They are known by many names in the UK depending upon what region you come from and there are some quaint and beautiful examples to be found. Here are a few to make you chuckle. Bibble-bug, Chizzle-ball, Fuzzy-pig, Granny-picker, Lockchest, Penny-mouse, Chisleps, Wood-pig, Chuggy-pig, Roly-poly, Snot, Cheese log, Tiddy-hog, Sow-pig, Grumpy-gravie and my favourite of all (from Devon) is the rather beautiful Gramfer-grig. I believe that gramfer means grandfather and grig is

thought to mean small – or dwarf-like. A

small grandfather it is then!

The woodlouse is actually a literary symbol. May I suggest that you take a trip to St Mary's Church in Shrewsbury where you will find amongst its many gems a plump little animal in a roundel of glass. It has above it the words *'sol in cancro'* – the sun in cancer - suggesting that it might be a replacement (for some reason) for a crab as the sign of the zodiac.

The common woodlouse *Porcellio scaber* means rough little pig and this tiny crustacean can be found in habitats right across the globe. There are some 5000 species and they all have 14 jointed limbs. They should not be confused with pill bugs that can roll up into a ball – most woodlice

cannot do this. Isn't evolution a remarkable process?

Woodlice have played a startling role in cures for illness down the centuries and were said to have medicinal properties. They were ground up and swallowed to cure stomach ache or simply eaten live. One believer said 'Mother's better – I gave her three sowpigs last night'. This practice was known up until the 1930's when Mr Albert Golding told how his grandfather in Bristol was treated for jaundice whilst a baby by grinding up woodlice in the mill.

I am sure that you all knew that students of woodlice refer to themselves as oniscologists and that there are around 40 outdoor species of woodlice in the UK, though only a

handful are common. Woodlice eat 5-10 per cent of their body weight every day and feast on rotting wood and even droppings.

I should like to give some excerpts from a woodlouse poet, Jean Kenward who pitied the creature's vulnerability.

Deep in a crumbling
darkness, crisply armoured
against attack,
grey woodlice are assembled, dry and silent
cushioned in cleft and crack

Cold, spherical
steel hard, they fold their tiny
bodies so tight
as to allow
no entry to the summer's
pervasive light

Only – at a brief
rising of the curtain –
In sudden, wild hysteria, they run
this way, and that

each one must brace
himself to bear
the bird infested morning
and the sun's face.

Rather amusingly in Wiltshire, children used to recite a chant when holding a woodlouse in the hand.

'Granfer Grig killed a pig
Hing un up in corners
Granfer cried and piggy died
And all the fun was over'

I wonder if children share such charming moments like this today. I am afraid that many that I speak to are completely ignorant, afraid or oblivious of the natural world and the weird and wonderful inhabitants around them. I work hard to remedy this.

For me, woodlice are strange and ancient creatures and as someone who has an interest in arachnology I can always find the elusive woodlouse spider *Dysdera crocata* lurking near to a colony of our 'tiddy-hog' friends. I don't know what name is attributed to the fear of woodlice (though Isopterophobia seems to be closest) but I do hope that the unfortunate woman mentioned at the start of my column can overcome it and take a shower soon.

As a footnote to this column I have just returned from a wonderful trip to Belize and Guatemala, and in one of the caves that I investigated I came across some gorgeous 'cave lice' which can be seen here feeding on bat guano. There were hundreds of bats just above my head as I took this shot. The lice look so primitive, and with little use for any kind of eyesight they just have black dots and I wonder how well these function, if at all.

Born in Birmingham, England Carl Portman has always followed the maxim 'interest is where you find it' and this certainly applies to natural history. He has bred endangered species of tarantula spiders, written two books on natural history travel, and lectures around middle England on animals and rainforests. Oddly he has a diploma in sexing juvenile theraphosid spiders, is an English Chess Federation County Chess Master, supports Aston Villa and has a strange addiction to Turkish Delight (covered in chocolate). Having worked for the Ministry of Defence for 30 years he now spends his time doing lecturing, chess coaching, some photography and management consulting.

He has spent time studying animals in the rainforests of Australia, Ecuador and Costa Rica searching for new and ever curious insects and arachnids and has a desire to find a new species somewhere in the world. His motto is 'Don't complain about the dark, light a few candles'.

He is married to Susan and lives in Oxfordshire. Their two Border collies, Darwin and Dickens keep them fit and ensure that there is never a dull moment in the household.

watcher of the skies

CORINNA DOWNES

The news over the past few months has been overshadowed by the continued poisoning or shooting of birds of prey, and it is welcome news that some perpetrators have been caught and sentenced. Unfortunately, however, like the barbaric slaughter of endangered birds in Malta, I fear that this practice will not stop until some of these birds disappear forever. And that would be a very sad thing, and would reiterate – if indeed it really needs any reiteration – that humans are the most malevolent species to ever inhabit this Earth. But on a brighter note, there have been a fair few rare vagrants to visit our shores since the last edition of this magazine was issued, so thankfully the skies are not devoid of vibrant feathered life quite yet.

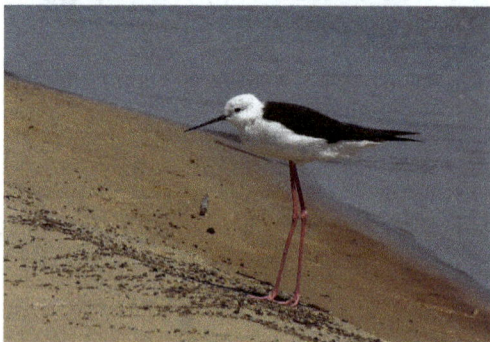

Six black-winged stilts, including four juveniles, were reported at Cavenham Pits in Suffolk from 19th July onwards, and it seems highly likely that the birds must have bred in the vicinity, perhaps on site. Five remained at Medmerry in West Sussex, while there was belated news of a single bird at Shapwick Heath in Somerset on 13th July.

Source:
http://www.birdguides.com/

A first Staffordshire record came in the form of a Pacific golden plover during July 2014. A summer-plumaged male was found on the RSPB Middleton Lakes reserve. This is an Asian species which breeds on tundra from Central Siberia east to western Alaska.

Source:
http://www.birdguides.com/

A pair of rare bee-eaters nested on the Isle of Wight for the first time on record in July. This is the third record of this bird successfully breeding in the UK in the last century, the first being in 2002 in Co. Durham and the second in 1955 in Sussex.

Source:
http://www.wildlifeextra.com/go/news/bee-eater-nesting-723.html#cr

In May 2014 a hoopoe turned up in Innerleithen and is thought to have overshot its migration to southern Europe. Hoopoes spend the winter in Africa before returning to southern parts of Europe for the summer. A small number "overshoot" their route and end up along the English south coast but they are rarely seen as far north as Scotland.

Source:
http://www.bbc.co.uk/news/uk-scotland-south-scotland-27508472

In August what looked to be a good candidate for a Scopoli's shearwater was seen, and if confirmed, it would be only the second record for Britain.

Source:

http://www.birdguides.com/

A marsh sandpiper was a first for Gloucestershire when it was seen in flooded fields at Splatt Bridge on 28th August.

Source:
http://www.birdguides.com/

During the summer a pair of glossy ibises was seen courting and building a nest at RSPB Frampton Marsh in Lincolnshire, and was believed to be the first recorded modern-day nesting attempt by this species in the UK.

Although they did not go on to raise young, this is a great coup for Frampton and a sign that the species could be on the verge of establishing a breeding population in Britain.

These birds usually live in the Mediterranean region and are rare visitors to the UK, although these visitations are on the rise making the bird more or less resident here.

Source:
http://www.birdguides.com

A masked shrike showed up at Kilnsea in East Yorkshire. This is one of the rarest representatives of its family on the British list (the species has yet to be recorded in Ireland).

There have been just two previous records, from far-flung Fife in 2004 and Scilly in 2006, and both in late autumn, so a mainland

bird in eastern England was always going to be welcome. If accepted, this will represent the third British record: the first being in Fife in 2004 and the second on Scilly in 2006.

Source:
http://www.birdguides.com

During October, a first winter eyebrowed thrush made an appearance on North Ronaldsay in Orkney, this being the first time for five years that one has been recorded there. This represents the 21st British record of this Far Eastern thrush. The Isles of Scilly boast seven records, six of which occurred between 1984 and 1993; this individual is the third for Orkney following the 2009 bird and another at Evie, Mainland, in late September 1984. While most records come from October, there are three spring records including a twitchable bird in Angus for three days in late May 1996.

Source:
http://www.birdguides.com

A scarlet tanager was seen on Barra on the Outer Hebrides, in woodland near Brevig on the afternoon of 6th October. This is the 11th record of this species for Britain and Ireland, but the first for Scotland. The species was last recorded in 2011, when first-winter males were in Cornwall and on Scilly. As is expected for this American vagrant, British and Irish records show a distinctly western bias. Adult males are striking red-and-black birds, but that plumage has only been seen once this side of the pond; the Barra bird, like most before it, has the subtler yellow-green plumage of the first-year male. The species breeds across the eastern half of the United States and migrates through Central America to north-western South America for the winter.

Source:
http://www.birdguides.com

A steppe grey shrike was recorded at Burnham Norton on the north Norfolk coast at the beginning of October. Steppe grey shrike is part of a confusing complex of grey shrikes of the genus Lanius, which some authorities have proposed should be split into as many as six species, but what all authorities agree on is that pallidirostris is one of the 'southern' forms; its usual range is in Central Asia, east of the Caspian Sea. This bird is the first for Norfolk, and the first in Britain since a bird in Shropshire in 2011.

Source:
http://www.birdguides.com

A flock of 47 spoonbills - the largest ever seen in Britain – was present in Poole Harbour, Dorset. This follows the species establishing a breeding colony in Norfolk in 2012. Paul Morton from the Birds of Poole Harbour charity said:

"To have 47 Spoonbills in the harbour is a fantastic sight, and goes to show how successful their breeding colonies are doing elsewhere. From [last] year's data, we suspect they have come from Holland or Belgium.

"For around 50 years Poole Harbour has

only ever had 2–3 Spoonbills during the winter, but this last decade has seen numbers grow year on year, as youngsters follow their parents back to their winter quarters. What makes this gathering even more special is that people at home, work and school can enjoy the spectacle live on the Brownsea Lagoon webcam 24 hours a day. The only question is: how long will it be before they start breeding here?"

Source:
http://www.dorsetwildlifetrust.org.uk/
spoonbills_on_Brownsea_Island.html

Although there have been over seventy combined British and Irish records, the yellow-billed cuckoo is a deceptively hard bird to see on our shores. Many past records involve dead or dying birds and the last to be remotely twitchable was as far back as 2000 at St Levan in Cornwall during October. Between then and 2011 there were no fewer than six occurrences but two were found dead, a third died in

care and three, though alive, made only brief appearances. The arrival of a bird at Porthgwarra in Cornwall was a welcome find. However, after its disappearance it was widely assumed that the cuckoo must have perished, or perhaps (hopefully) had taken advantage of the conditions and moved on.

Source:
http://www.birdguides.com

During November, Britain's third Eastern crowned warbler was discovered at the Hunley Hotel and Golf Club in Cleveland. This bird has only recently established itself as a rare but expected vagrant to Western Europe, with seven records since the bird in Norway in 2002 - before that there was only one Western Palearctic record, from Heligoland in Germany in October 1843.

Source:
http://www.express.co.uk/news/nature/530411/Third-
warbler-Cleveland-Golf-club-twitchers

There have been 229 previous records of little bustards in Britain, but it's been over twelve years since the last one was recorded here: a one-day male on St Agnes, Scilly, in 2002, and a bird on The Lizard in Cornwall, in late October 1996 marked the last twitchable bird on our shores. Therefore, despite the abundance of records (almost all of them from before 1970) this is very much

a contemporary mega-rarity and the arrival of a bird at West Bexington in Dorset, on 18 November was a great find. During the Christmas and New Year period a little bustard was recorded in East Yorkshire in a field near Fraisthorpe, and became the first twitchable of its kind in Britain since late October 1996. Though the little bustard breeds as close as Iberia and southern France, it is widely accepted that many of our vagrants will originate from eastern populations and it seems likely that, given the freezing conditions on the Continent over Christmas, the Yorkshire bird (as well as one in Sussex) likely came westwards before arriving on British shores. Little bustard breed widely across the steppes and grasslands of eastern Europe eastwards throughout central Asia, wintering as far south as the Middle East. Its distribution across southern Europe is much more fragmented, although healthy populations do still occur across Iberia and also in parts of France.

Source:
http://www.birdguides.com

An Isabelline wheatear was found on the beach at Seaton Snook in Cleveland during November last year. These birds breed from Greece, Bulgaria and Turkey eastwards through Central Asia as far as China, wintering south to India, throughout the Arabian Peninsula and along the Sahel belt of Africa. There are just 33 previous accepted records here, although the species has become more frequent in recent times, with 16 of these occurring since the turn of the century.

Source:

http://www.birdguides.com
At the beginning of December, Wales confirmed its first Caspian gull; a first-winter bird at Gresford Flash in Clwyd. Remarkably, the bird was joined by a second – this time an adult – on 6-7th. The Caspian gull is a regular winter visitor to areas of the South East, Midlands, East Anglia and as far north as Cleveland, though records further north and west of here are few and far between. That said, the long wait for a twitchable Welsh bird has been nothing short of amazing given how regularly birds are seen in the likes of Shropshire and Cheshire. Another first-winter was subsequently seen and photographed at Cosmeston Lakes, Glamorgan, on 8th.

Source:
http://www.birdguides.com

During December a Blyth's pipit was discovered inland near Wakefield in West Yorkshire, in a rough field on an industrial estate adjacent to Calder Wetlands. This

was a first for Yorkshire and the third individual of the year following those on the Isles of Scilly and Pembrokeshire.

Source:
http://www.birdguides.com

During January 2015, Ireland's first confirmed black scoter, a drake found off Rossbeigh, Co Kerry, came as a surprise,

given that there have been several British records as well as numerous individuals in the likes of Denmark, Sweden, Poland and so on.

Source:
http://www.birdguides.com

For those of you not aware, as well as this column in *Animals & Men,* Corinna writes a daily Fortean bird blog which can be found as part of the CFZ Blog Network, but also as a stand alone site at:

http://
cfzwatcheroftheskies.blogspot.com/

Tasmanian Wolf at the Door

Richard Freeman

The rain was battering the tent so hard it felt more like hail. Even in two sleeping bags and fully clothed I could feel the biting cold. I have a morbid loathing of cold. This was Australia. It was supposed to be hot and dry, a sun-kissed paradise. Trust me to come to the one part of Australia that had weather as bad as Britain. It was early November in Tasmania. Over there it was late spring but the weather had been uncommonly inclement. My old friend Tony Healy, who had been here several weeks before me, doing ground work, had spoken of horizontal snow. But I was here for a good reason. I'd waited a long, long time to visit Tasmania. I was on the track

of a living legend, an icon of conservation and, I believe, a true survivor.

Of all the world's cryptids, the most likely to exist is the enigmatic and beautiful creature known as the thylacine. This flesh eating marsupial is one of the most spectacular examples of convergent evolution, where two different species, often on opposite sides of the world, bear a remarkable resemblance to one another due to each inhabiting similar ecological niches. The thylacine (*Thylacinus cynocephalus*) is also known as the Tasmanian wolf or Tasmanian tiger, and is convergent with the placental wolf. The animal bears a striking resemblance to a

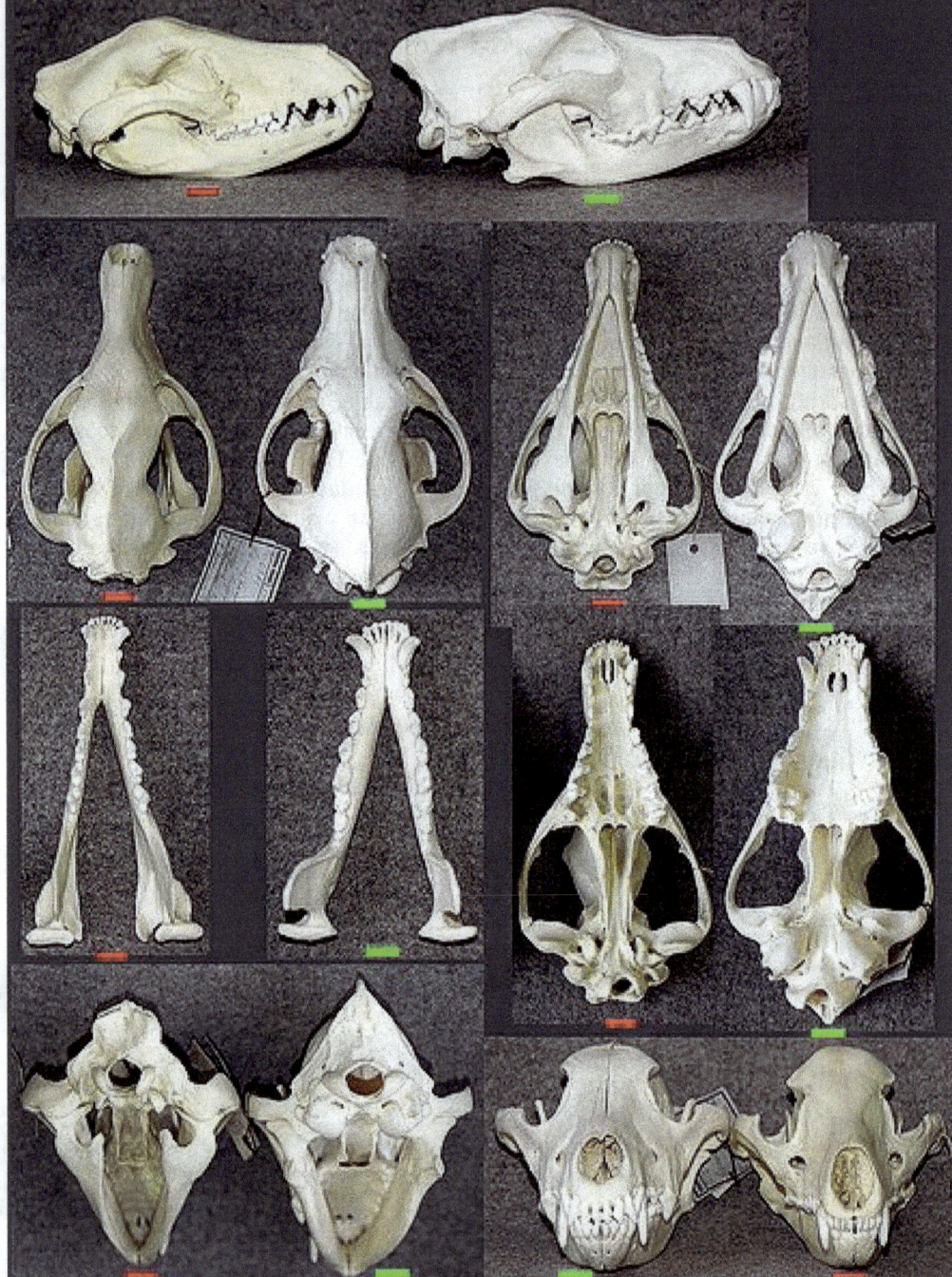

Copy of a skull of *Thylacinus cynocephalus* and original skull of *Canis lupus*, Museum Wiesbaden Naturhistorische Landessammlung, Germany Fritz Geller-Grimm

wolf or dog but with stripes along its hind quarters. Of course it is not related to the wolf or the tiger. Neither should it be confused with the Tasmanian devil (*Sarcophilus harrisii*) a superficially badger-like flesh-eating marsupial, or the spotted or tiger quoll (*Dasyurus maculatus*), a native cat-like marsupial predator. Both sexes have a backwards facing pouch. In females it is used to nurture and protect developing young and in males to protect the sex organs as it runs through vegetation after prey. The skull of the animal has a gape far wider than that of a wolf or dog. The thylacine's dental formula are different to a wolf's. It bore four incisors and four molars in each quadrant of the jaw as opposed to only three of each in true canids. The thylacine has a more powerful bite than a wolf but the skull was less adapted to holding struggling prey. This suggests a different hunting strategy. Where as pack-hunting wolves would use numbers to pull down prey and worry it to death, thylacines may kill small prey animals with one bite, and with larger victims inflict a bite then let them bleed to death. It is not as well adapted to fast running as a wolf but seems to have more stamina for pursuit over long distances.

The thylacine is the largest marsupial predator of recent times and has a lineage that reaches back to the Miocene epoch. Thylacines were once found across mainland Australia and New Guinea as well as Tasmania. Standard thinking would have us believe that the species died out on the mainland around two thousand years ago, perhaps from diseases transmitted by the introduced dingo. However, sightings persist in both Australia and New Guinea until the present day.

When white settlers first colonised Tasmania in 1803 they began an act of ecological genocide. The largest broad-leafed trees on earth, the giant mountain ashes, were cut down. The Tasmanian black emus were hunted into extinction by the 1830s. The Tasmanian Aboriginal populations were decimated by hunting and disease. Their culture has almost entirely vanished, and only vestiges remain.

Areas of forest were cut down to allow the grazing of sheep. The Tasmanian wolf was an inconvenience for sheep farmers. Doubtless the creature did indeed kill some sheep. Slow-moving, placid targets are hard for predators to resist, but the claims of predation by some sheep farmers were on such a scale as to be physically impossible.

Like most politicians everywhere and at every time, the Tasmanian government were self-serving cowards with knee jerk reactions. To be seen as doing *something* they set up bounties on thylacines from 1830 to 1909. The bounty was set at £1 per head. Between the above dates 2,184 bounties were paid. By the 1920s the thylacine had become scarce in the wild. A thylacine was shot by Wilf Batty at Mawbanna in 1930. Elias Churchill trapped one alive in the Florentine Valley in 1933.

Many specimens were caught for zoos around the world including London, but no concerted attempt was made to captive breed them. The last captive animal died on September the 7th in 1936, apparently from cold as it had been locked out of its sleeping quarters.

Since the date of the Tasmanian wolf's

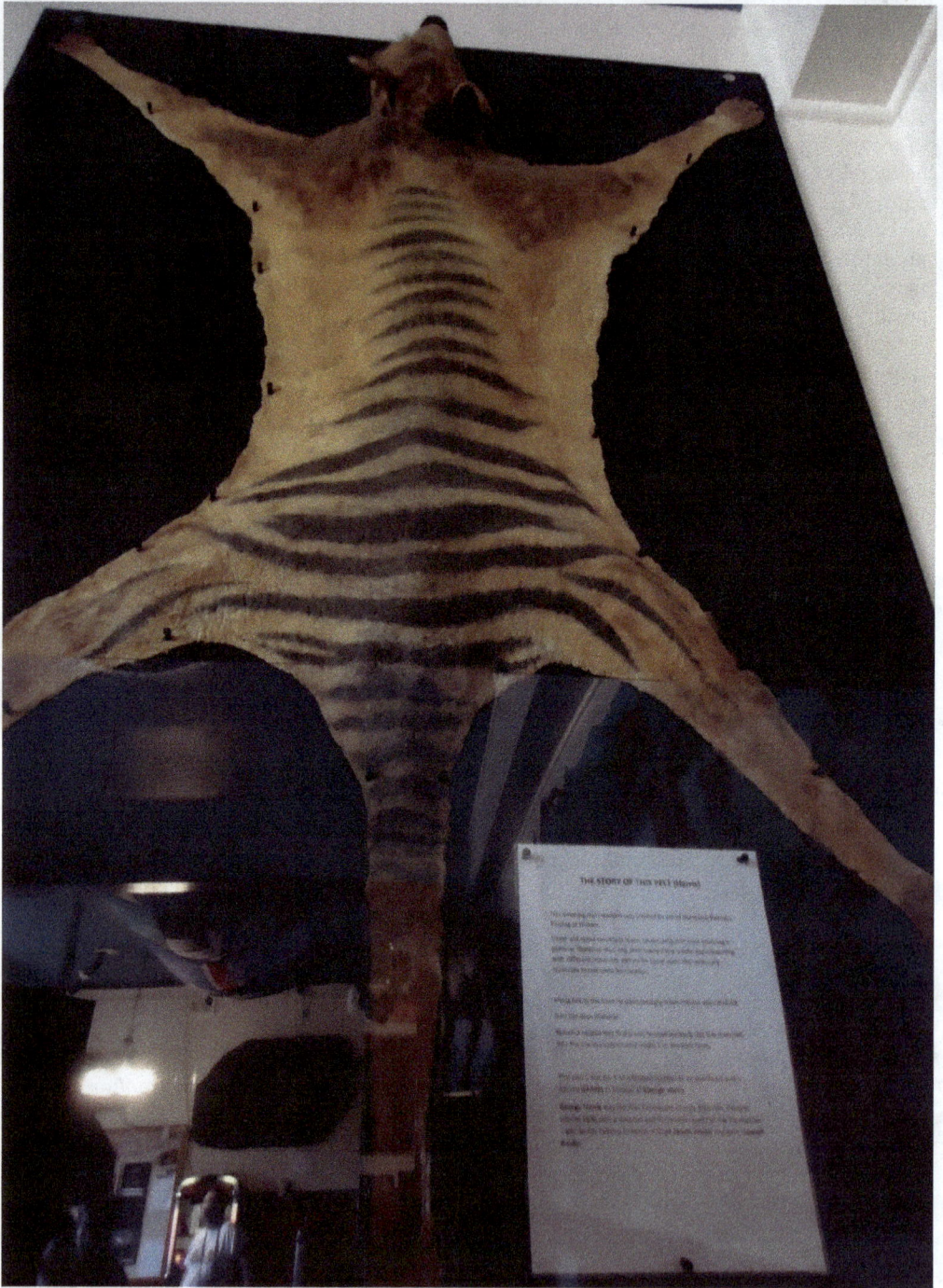

THE STORY OF THIS PELT (Skin)

official extinction there have been more than 4000 reported sightings. These come not just from laymen but also from some very credible witnesses including zoologist Hans Naarding, who in 1982 observed a large male thylacine near the Arthur River in the state's northwest.

The creature's continued survival has even been predicted by computer programme. Professor Henry Nix of the Australian University's Centre for Resource and Environmental Studies developed a computer programme called BIOCLIM. A research tool, BIOCLIM matched what was known about a species habits and preferences and geographical areas. It matched the two up and predicted where, within a given area, the target species was most likely to be found. Nix applied this to the thylacine and the BIOCLIM programme. There was an almost perfect match to where the programme predicted the animals would be if they had survived and the areas where sightings were being made. Nix concluded that people were really seeing thylacines.

The following factors should also be noted. Firstly, island populations frequently require a much smaller base number to create a viable breeding group. For example, the naturally low genetic diversity of the Tasmanian devil means that a viable base population could be as low as 25 individuals. Professor Nix thought that as many as 1000 thylacines may still exist island-wide. Secondly, the south west of Tasmania was never settled save for a handful of tin miners and fishermen at Port Davey. The area itself produced no thylacines during the bounty period. The area is not ideal for the animal but we know that creatures under pressure can retreat to and indeed thrive in less than

perfect conditions. A good example is the recently discovered population of Bengal tigers *(Panthera tigris tigris)* living in the high Himalayan mountains in Bhutan at an altitude of up to 11500 feet, far above their normal range. Therefore, it is quite possible that thylacine populations moved into the southwest during the bounty years and remained unmolested. Eventually these would have re-colonised other areas of the island. Today most reports come from the northeast and west of Tasmania, and the west coast.

I had always wanted to go in search of the thylacine. Back in the '90s as a student I spent some time planning an expedition only to be lied to and have my research stolen by a UK film company. Something that I am still bitter and angry about.

When the Centre for Fortean Zoology's Australian representatives Rebecca Lang and Mike Williams came up with the idea of an official CFZ expedition to Tasmania, I was thrilled to be on board. Mike and Rebecca were to be joined by another old friend of the CFZ, veteran Australian cryptozoologist, Tony Healy. Tony had spent a lifetime on the track of unknown animals all across the world. Also joining us would be Tania Poole, a CFZ member and researcher who had joined us at the Weird Weekend, the CFZ's annual convention and the *Fortean Times* Unconvention on a number of occasions. Rebecca's friend Hannah Jenkins would round off the Australian team.

The British contingent left Heathrow on October the 31st. It felt truly special to be searching for the creature so iconic that the CFZ had adopted it as its logo and totem animal.

I'm used to long flights, but the trip to

Tasmania was something else. We stopped in the Middle East, Borneo and Melbourne before reaching Launceston in Hobart more than 24 hours later. We were met by our Australian friends and in no time we were driving to Launceston. Tasmania is alive with wildlife and on our first relatively short journey we saw wombats, wallabies and an echidna.

Launceston, Tasmania's second city, is the size of a medium town in the UK. It has an old-world feel about it with its colonial era buildings and houses. The Tasmanian wolf is everywhere, on car registrations, in shop signs and council logos. The creature is very much alive in the island's iconography. It even appears as supporters on the Tasmanian coat of arms.

Moving on inland we reached the small town of Mole Creek in central Tasmania. We had been booked into the Mole Creek Hotel, home of the famous Tasmanian Tiger Bar. The hotel itself has a sort of old fashioned charm with a 1950's feel to it. I felt very comfortable there and it had a lovely atmosphere. The Tasmanian Tiger Bar is like a small museum filled with thylacine memorabilia. There are paintings, sculptures and a wall full of framed newspaper reports of sightings. They even serve Tasmanian Tiger Ale, a very tasty pale ale of which I imbibed several pints.

The landlord, a charming man called Doug Westbrook, was good enough to give us an interview. He showed us alleged droppings (desiccated and in a jar) and a number of prints. The prints did indeed match those of a thylacine rather than a dog, wombat, fox or any other animal. Doug himself had never seen the Tasmanian wolf, but his wife had.

Doug's wife Ramona was lucky enough to see a thylacine in 1997. She encountered it on the road from Mole Creek to Paradise, close to the spot where a bridge crosses the Minnow River, about 18 km from Mole Creek. As it was about 6.15 on a summer's evening - full daylight – she had a very good look at the animal.

As she drove past it – too frightened to stop – it stopped beside the road and looked at her. What struck her most were its "big dark eyes – strange eyes – dark, like they had eye-liner around them."

Doug remembers her rushing back home, extremely excited, and adamant she'd seen a tiger. A few other people have reported sightings on the same stretch of road. A young French backpacker, who was working in Mole Creek, saw one in 2010 close to the location of Ramona's encounter. That one also crossed the road ahead of her car. She described its striped rump, stiff tail and strange gait. She used the phrase 'like a dog with a broken back' to describe the way it moved.

We visited the Trowunna Wildlife Park close by. As well as birds, reptiles, kangaroos, wombats and echidnas, the collection included the island's other marsupial predators - spotted quolls and Tasmanian devils. Up close the resemblance of the quoll's face to the thylacine's is striking. Both have dark eyes, rounded ears and a dog-like snout. There the resemblance ends. The body and tail look more like a stout cat with brown fur and cream spots. The bulkier Tasmanian devil looks more like a hybrid of bull terrier and badger. The devils are currently beset by a form of transmittable cancer that affects the face of the animal. First seen in the mid-1990s the disease causes huge facial tumours that lead to death. Devil facial tumour disease has caused a population crash of 50 per cent. Trowunna maintain a large, healthy breeding population in captivity as a safeguard against extinction in the wild.

We travelled in two magnificent Toyota Landcruisers, kindly loaned to us by the company for the expedition, and Tony Healy's trusty old van. On the road Tony unveiled his maps like he was revealing a wizard's spell books. They were dotted with notes and annotations in astonishing detail. Arrows and dots pointed out locations and dates of sightings not only of Tasmanian wolves but of bunyips, sea serpents, yowies and even ghosts.

We headed out to Cradle Mt - Lake St Clair National Park and the Wilderness Gallery where there was an impressive and shocking exhibition on the thylacine. It featured a thylacine skeleton, pelts, skulls and a reconstruction of an old trapper's hut from the 19th Century. On display was a book logging the captures and killings of thylacines from the bounty era. It covered the late 19th and early 20th Century. It felt odd to be actually touching the book, with its original notations of specimens, locations and payments. The numbers being brought in dropped sharply in the 20th Century.

The main exhibit was the Tiger Buggy Rug, an object both appalling and fascinating; a carpet made from the hides of eight thylacines. A film on a loop showed the last captive thylacine wandering around its barren enclosure at Hobart Zoo. I've seen the film many times before, but it was intercut with other, older, rarer clips of captive animals.

Annoyingly the whole exhibition focused on thylacine extinction. There was not one word about thylacine survival or any of the 4000 plus sightings since 1936.

Prior to embarking on the expedition proper, we had all agreed to keep the focus area, and the identity of witnesses, a secret. Therefore I will not reveal exactly where we did our work or who we interviewed at this stage, other than that it was in the north and north west of the island.

On the way we saw much wildlife including another echidna and the ubiquitous Tasmanian native hen. The birds, which are actually flightless rails, are found just about anywhere there is water.

We set up camp without delay. Tony was sleeping in his van, the rest of us in tents. CFZ stalwarts Jon Hare and Chris Clark had two small tents of their own. Mike and Rebecca shared a larger one and taxidermist Jon McGowan, Tania Poole and myself shared Tania's huge tent, which she had inherited from an old flatmate.

The first campsite had its own residents. One was a noisy brush tailed possum that disturbed the night with vocalisations that one would not believe such a small and endearing animal could make. The second was a black currawong, a yellowed eyed, crow-like bird always on the lookout for scraps.

The forest floor around the camp was studded with what looked like huge worm casts. We dug down into the earth to try and find what we initially believed to be the massive worms that had created them. We had no luck.

That night we conducted the first of our night drives. As 50% of all tiger sightings occur when the creatures run across in front of moving vehicles, we had cameras mounted on the windscreens of the Toyota Landcruisers and Tony's van. These were left running throughout the drives.

The area we were searching in was remote. It was heavily forested and well away from main roads. We followed old logging tracks, some unused for years. The forest was thick with wildlife, all of which would make fine prey for the Tasmanian wolf. On the first night alone we saw Bennett's wallabies, red-bellied pademelons (a form of small wallaby), wombats and a Tasmanian spotted owl.

Despite being the beginning of summer in Tasmania it was still bitingly cold at night.

Next day we took a trek to a local lake; a large, water-filled sinkhole. Around this particular lake people have claimed to have heard the distinctive call of the thylacine at night. It is said to be a high pitched yap in three parts 'yip-yip-yip'. It's said to be quite distinct from all other native animals and quite unlike a fox's.

We set up some camera traps, sensitive to both heat and motion. These we baited with leftover chicken, cat food and bacon jerky. We set up other cameras along a long closed and barricaded road reasoning that this would be doubly undisturbed. We searched for road kill to use as further bait but found none.

Along another one of these logging spurs we found some large droppings. They were transparently those of a carnivore, containing as they did bone shards and hair. They seemed too large to be from a devil or quoll and too remote to be from a dog. We carefully preserved them in a solution of 70 per cent methylated spirits and 30 per cent water. Thylacines were reported in this area in 1994 and 1996.

The area had several small rivers running through it and we came upon a bridge that was so rotten it would not hold our vehicles. We carried on, on foot.

Later we visited an open area of button grass. As darkness fell we took watch. It was bitingly cold and perfectly still. Nothing moved and we saw and heard nothing.

We came upon a dead chicken and hung it beside one of the old logging spurs with another camera trap facing it. Another night drive turned up more wombats, possums, pademelons and wallabies.

The following day Tony took his van into town for a service. In the garage he met a woman whose father had seen a thylacine twice in the general area we were in, back in the 1980s.

Jon McGowan, Chris, Mike and I searched for snakes with little success. After nightfall we went out on foot spotlighting for animals and again saw much fauna.

The next morning we drove for a few hours to talk to a man who had seen a thylacine in 1996 and he was good enough to grant us an interview.

He had been driving between Hobart and Launceston when a large thylacine had rushed across the road. Once again the stiff looking, striped hind quarters were emphasised. It was the size and general shape of a large dog and he pointed out his own dog, a large crossbreed, as a good comparison. He also mentioned a stiffly held, thick tail. He noticed how, after crossing the road, the thylacine was moving up and down looking for a way through a wire mesh fence. The creature ran back across the road. Another car behind him had slowed down and also saw the animal. The occupants of the second car must have reported the sighting as a story about it subsequently appeared in the *Launceston Examiner.*

Often on our expeditions we turn up information on cryptids other than the one we were are actually looking for. This trip was no exception. We were talking to this witness about Tasmanian wildlife in general and he mentioned that when he first moved into the area that something had been killing his chickens. He shot the offending predator and it turned out to be a spotted quoll, but one of mind-boggling size. This animal is usually around seven pounds in weight and around three feet long. He said the animal he shot was the size of a cattle dog. This breed of herding dog is slightly larger than a border collie, weighing up to 49 pounds. He indicated the tail and body length by raising his hand from the ground to just above his shoulder. The witness was a tall man, over six feet, so the length of the giant quoll would be over five feet. He commented on

the thickness of the animal's neck. In comparison, the thylacine averaged on six feet in total length with some larger individuals ranging from seven to nine and a half feet long. At the time he had no idea of the value of such a specimen and threw it away. This occurred in 1964.

Later that day we spoke to another witness who saw a large, dog-like animal run across the road in front of his vehicle in 2008. He noted the stiffly held tail but could not recall stripes. However, there are other records of stripeless thylacines. Most striped animals have variation of their marking. For example, there are tigers with very faint stripes.

Then in 2010 he and a friend saw another thylacine a few miles from his first sighting. This time he did see the stripes as well as the stiff tail and odd hindquarters as it ran away into the bush. ˙ The following year he was approached by a logger who asked him if he had ever seen a thylacine. He said he had, and the logger confessed he had too, in broad daylight and in the same area. The man had been on foot and walking along the logging spur when he saw the thylacine.

Our informant had also heard the distinctive call of the Tasmanian wolf on several occasions. It was distinct from a fox's and more spasmodic. One of the places he had heard it was at a small, remote airstrip on which wallabies grazed at night.

Later that day Jon McGowan came across a road kill bandicoot. He cooked and ate some of the creature. Back in England Jon, who works at the Bournemouth Society of Natural Science, lives almost entirely on road kill, feeding his guests on badgers,

Bushnell   11-06-2013   09:57:16

foxes and other strange delicacies. He offered me some smoked bandicoot flesh but it smelled a little high for me.

The next day a wildlife guide and a couple of elderly tourists crossed our path. The man told us that the creatures that caused what we thought were huge worm casts were in fact burrowing crayfish. They have networks of burrows and shafts running up from the water table, and play a major role in soil turnover, drainage and aeration. The guide said that he too had heard the distinctive yip-yip-yip of the thylacine on two occasions.

The following day was Jon McGowan's birthday and he received not only a large cake, but also the gift of some road kill – a Bennett's wallaby. The creature had been found by some of the team when they had driven to a small town that morning. Jon gleefully skinned and butchered the creature.

We used some of the meat to re-bait a number of the game cameras. We took the opportunity to look at the images we had captured so far. The camera from the lake showed a tawny frogmouth (a nocturnal, nightjar-like bird), Tasmanian devils and a spotted quoll. The camera from the logging road showed devils and a feral cat. Jon Hare had spoken to a woman at a café in a town we had passed through who had seen a Tasmanian wolf crossing the road close to the area where we were camped. Unfortunately he did not get any further

details.

In the same town Tony interviewed a mechanic who'd seen a thylacine at very close range in 1982, during a time of severe drought and large bush fires. He and a mate, both then 18 years old, encountered it at about 10.30 one night, as they were riding a motorbike about eight kilometres from town.

The animal, the size of a large dog, was on the road just past a bridge over a small creek. They saw immediately that it was "something odd" and realised practically straight away that it was a thylacine.

They slowed, rode past it, then stopped and turned the bike around so that the creature, then only about seven metres away, was illuminated by the headlight.

Although the stripes were only faint, the tail and rear quarters were very distinctive, and could not have been those of a dog. The weird hindquarters "couldn't have been faked".

"It sort of swayed its head and body" and began to move towards them. That was enough for the teenagers: they took off.

He and his mate, he said, "maintain to this day it was a tiger." He thinks it may have been driven from its usual haunts in wilder country by the bush fires, and was hunting along the creek.

That night we barbecued wallaby. Even after using the head, tail, feet and innards as bait there was a large amount of meat left. It proved to be palatable, if a little tasteless. The night brought a savage rainstorm.

We visited another remote area where we had been told a witness had heard the thylacine's call, which was frequented by wallabies, wombats and kangaroos judging by the vast amount of scat. We also visited some old logging roads where thylacine sightings had been reported. They were exceedingly overgrown. It was obvious no one had been along them in some time. Tony and I saw a magnificent and deadly Tasmanian tiger snake slithering across the road in front of us. We leapt from Tony's van, but were too late to catch up with the reptile.

We returned to the area later that night to stake it out. We heard a wallaby give a thumping alarm in the manner of an overgrown rabbit. We also heard several owls. The night sky was punctuated by shooting stars and the rays of the Aurora Australis or 'southern lights'. Any thylacines lurking in the shadows remained silent.

Next day we encountered another witness who told us he had seen a thylacine some way from where we were searching. He was walking on a path off the river, and he saw the animal crossing the path ahead of him in broad daylight. He had been struck by the stripes on the flank. This occurred in 1988.

Later we took down the camera traps and reset them with fresh bait at another area and along the roads and hills around it. Rebecca spent the night at the area where

zoologist Hans Naarding had an excellent view of a large male thylacine in 1982. After spending a freezing night in one of the Landcruisers she saw nothing.

Next day we drove to Mawbanna where one of the last known wild thylacines was shot by Wilf Batty in 1930. Rumour has it that six more were caught alive in the general area in the 1930s, and one in the 1960s. These days the place looked quite unsuitable, being mostly cleared farmland. That night at camp Jon McGowan returned

in a state of excitement after wandering in the surrounding forest. Opening up his hand he showed us his prize with all the enthusiasm of a schoolboy. It was only a deadly funnel web spider with venom quite capable of killing a person. Rebecca, being an arachnophobe, was appalled as Jon played with the huge arachnid as if it were a pet mouse!

We attempted to drive to a location a few hours north where thylacine tracks had recently been found. However, the sat-navs malfunctioned and we got lost. We ended up in the town of Waratah and had lunch at the excellent Bischoff Hotel, a magnificent building dating to the 1900s. Inside was a preserved specimen of the Tasmanian giant crayfish. It was as large as a big marine lobster. The owner told us of a local family who saw a thylacine crossing the road in front of their car near Rapid River in the 1970s. Apparently they disliked talking about it.

At the small local museum a 1970s newspaper was on show. It had a double page spread about the Tasmanian wolf, and detailed several sightings.

On the way back we stopped to explore more remote logging roads. On one we found more large droppings from a carnivore and took them as samples. Close to camp a spotted quoll bounded across the road in front of us and we caught it on camera along with several Tasmanian devils.

Rain marred the following day. Exploring the woods around the camp we discovered a cave. We rigged up some ropes and one by one lowered ourselves down into it. Spindly cave spiders with a leg span like a human hand lurked in the cave, apparently rare and found only on the island and in this part of Tasmania. Some sat upon egg sacks as large as hen's eggs. *Hikmania troglodytes* is the biggest spider on the island and belongs to a primitive group that is ancestral to modern spiders. Its closest relatives live in Chile and China.

Australian TV channel ABC wanted to do an interview with us. As we were nearing the end of our expedition we agreed. We met them in a small seaside town. Whilst waiting in a café to meet them I picked up a magazine and found an article written in it that was an almost word for word rip off of one I had written years before for the CFZ journal *Animals & Men*. It was about the creature known as the Gurt Dog of Ennerdale that terrorised the British Lake District in 1816. Its description and habits

recalled a Tasmanian wolf and I theorised that the creature had escaped from one of the horse drawn, travelling menageries that were popular at the time.

Finally they arrived. There was a likeable cameraman in his '50s and a young, somewhat pushy female presenter. The presenter wanted to film us discussing the expedition over a drink in a pub. We went to a pleasant pub overlooking the ocean and she ordered us a round of drinks. We were duly filmed talking about the thylacine and our trip. Then they wanted to film us 'setting up' our cameras. Of course we were not going to let on to the real location of our expedition. We would just recreate what we did in some nearby bushland. As we began to leave, we found out that the girl had left without paying for our drinks. We had to pay for them ourselves! We were quite annoyed by this, but things got worse. She was obsessed by bigfoot and kept saying "you have hunted for bigfoot haven't you?" I repeatedly told her that I had hunted for the yeti, the almasty and orang-pendek but never for bigfoot, to which she said "I've heard you have hunted bigfoot". Again I reiterated that I had not and neither had the other team members. She seemed like an ill-mannered child. We were filmed setting up the camera traps and interviewed about the expedition. When the piece was finally transmitted the presenter said that we had previously hunted for bigfoot despite what I had told her.

Later we found a recently dead road kill spotted quoll. Jon McGowan later skinned the animal and cooked it. The meat was succulent and far better than that of the wallaby.

While Jon Hare, Chris Clark, Mike

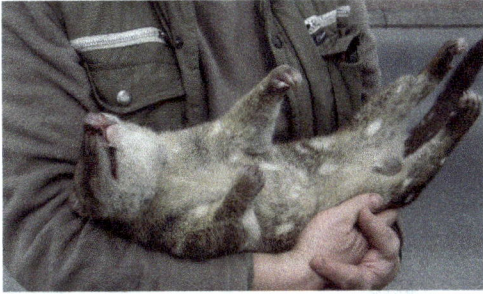

Williams, Jon McGowan and I had been enduring the witterings of the ABC reporter, Rebecca and Hannah had managed to get a look around a little museum in a nearby town and found an interesting photo of a stripeless thylacine.

On our final full day in Tasmania we had a remarkable stroke of luck. Tony and Mike were talking to some folk in a café in another small town we visited. One of these men thought the thylacine could well still be about. He said that if he had the money he would search south of the Arthur River. He related hearing that thylacines were fond of eating birds and that the ones in London Zoo caught pigeons. He also said he had been told of them hunting seabirds on beaches. I myself have read of them catching sparrows in captivity.

We returned to England, and our Antipodean colleagues returned to the mainland. The samples were sent off to Copenhagen University to be examined by our old friend zoologist Lars Thomas. Perhaps unsurprisingly they turned out to be those of a Tasmanian devil. It must have been a specimen of huge size.

We have already made plans to return in February of 2015. As we have done with the orang-pendek of Sumatra, we intend to keep returning to Tasmania on a series of expeditions. The small population and vast wilderness have convinced me more than ever of the Tasmanian wolf's continued existence. Tasmania is the size of Ireland, but its population is less than half a million and most of these live in Hobart or Launceston.

Interestingly, shortly after my return I came upon a very interesting book with a striking passage about the Tasmanian wolf. *The Ocean Inside* was written by Philip Hoare, Visiting Fellow at the University of Southampton and published by Fourth Estate in 2013. It consists of a number of essays on the world's oceans and on certain islands. In the chapter on Tasmania, the author writes extensively on the thylacine and modern day sightings. He finished the chapter with these words:

"What I do know is that in one institution I visit, a curator lets slip a quickly retracted remark, telling me it is not their secret to reveal. It is clear from what this person says, or does not say, that this strange half-life limbo of an animal which may or may not exist may soon be resolved, in its favour. That history is about to be reversed. That the thylacine is no longer extinct.

If it ever was."

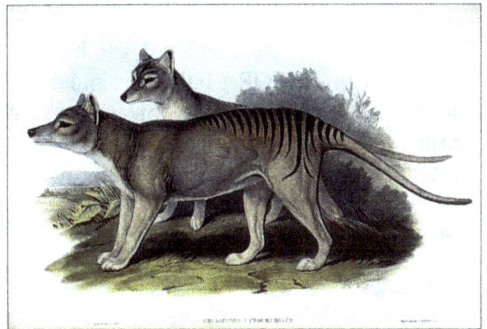

I originally came up with the idea of the Weird Weekend in 1999 after my friend and sometime colleague Nigel Wright and I had travelled around the Kingdom appearing at various conferences. Basically it was like being back at primary school again. When one has been invited to a certain amount of birthday parties thrown by other children, one has to badger one's parents into throwing one for you, whether or not you actually want to. It is sheer good manners. And Nige and I felt much the same about our own conference. So in May 2000 we held our own event for the first time and lost seventy-five quid doing it.

We had no intention of ever doing another one, but Judith from BUFORA bullied us into doing one, and - basically without us meaning to - it became an annual event. Then in 2005 I left Exeter never to return. I inherited my old family home in a little village in North Devon, and took both the CFZ and The Weird Weekend with me. From 2006 to 2013 we held the Weird Weekend at the Community Centre in Woolfardisworthy, the village where I live. Then in early 2014 things began to change. The Village Hall committee put the price up to what I thought was a ridiculous level. Initially they said that it was because the money was being raised for a charity from another village, but then they changed their mind and said that it was just in line with their new policies.

I heard from other village organisations that they, too, were being charged far more than they had expected for events. I wrote to the committee proffering the viewpoint that in my opinion, a community resource was perfectly justified in charging commercial rates for weddings, 21st birthday parties etc, but that I

The Weird Weekend 2014

believed that they had a duty to subsidise community events, especially ones that brought money into the village, and raised money for local charities.

They didn't agree, and so I took the event to the neighbouring village of Hartland. This caused a fair amount of rancour within the village, some of which I managed to sort out and some of which I didn't, but as I have made a career out of irritating people as some sort of an art form, I don't really mind. These days I keep myself very much to myself, and have no particular desire to socialise wildly so the fact that some of my fellow villagers dislike me because of some imagined slight against the village, matters about as much to me as the fact that some of my fellow villagers dislike the stance that I/we have taken on fox hunting and the badger cull. I truly don't care.

Because there had been a big falling out

between the ladies who for some years had been doing the Weird Weekend catering in aid of village charities, I had already decided that monies raised by the 2014 event should go towards the Small School in Hartland.

The Small School was founded in 1982 by Satish Kumar and other parents living in an isolated rural community in an economically-deprived area of South West England.   The nearest state secondary school, with almost 2,000 students, was 13 miles away, involving 2 hours travelling a day by bus. This pioneering group, most of whose children had been educated in small village primary schools, wished to show that secondary education, too, could be modelled on the family, rather than the factory, and based in the local community.

"The school is in the centre of Hartland in the old church hall and at the heart of the community. At the rear of the school we have a

vegetable garden that is maintained by the students and the food produced is used for the cooked lunches. As a school we aspire to a greener future and we are constantly looking at ways to be more environmentally friendly. As a school we recycle and source all our produce (if it's not already growing in the garden) from the local farm shop in Hartland. By doing this we are not only supporting local businesses but also cutting down on food miles.

The school serves vegetarian food and other dietary requirements are also catered for. A different parent volunteers to cook the lunch each day and a rota of students help out in the kitchen too. All students attend a Level 2 Food Safety course in order to prepare for the kitchen work. Students also take responsibility for the cleaning of the buildings at the end of the day."

It was my dear secretary Andrea Rider who first suggested that we hold the event at The Small School, and whilst it is a much smaller venue than the Woolfardisworthy Community Hall, it is cosier, and saw a return to the proper community feel that the event used to have when speakers slept in helper's spare rooms, and everyone collaborated on the food and drink. I was very happy with the event as a whole, and feel that it will probably continue for the foreseeable future, now I have a whole bunch of new helpers who can help the event evolve.

The 2014 Weird Weekend was considerably different to the last few years at the Woolfardisworthy Community Centre; for one thing some of the content was more overtly political. Wally Dean, the current custodian of the ashes of Wally Hope was one of the guests, for example. I salved my Fortean conscience by explaining to everyone (and myself) how the fact that Wally's mortal remains were

carried from sacred site to sacred site and from festival to festival, and that he even had miracles attributed to him, made him some sort of modern saint, or at least a holy relic, but the truth was that I have always been outraged by the story of Hope (real name Phil Russell) and how he was silenced, and effectively murdered by the British establishment for the crime of telling people to take their clothes off and dance in the communal sunshine.

It was interesting how this story polarised the audience. Some people couldn't see past the drug references and believed that he had "got what was coming to him" because of his proselytising about drug use. Others, like me, didn't care about the subject of the drugs, or at least weren't shocked by the subject, and saw reflections of the tragic story in their own lives and those of their friends.

Another highlight of the weekend was a lecture by my elder stepdaughter Shoshannah. She joined the bill at the very last moment, as advertised speaker Lee Walker had to pull out the day before because of the sudden death of his father. Shoshannah offered to do a talk about the veterinary implications of various quasi-cryptozoological phenomena, and I gladly accepted. Now, I don't know how to say this next bit without sounding appallingly insulting to the dear girl, but I shall do my best.

I have known her for just a few months under ten years now, and in July this year I will have been her stepfather for eight years. I knew she was intelligent and articulate, and that she knew her stuff, but I had no idea about her presentational skills. In fact, if you had asked me, I would probably have politely said that I was sure she would be fine, whilst privately thinking that her presentation might be a little hesitant in places.

WRONG!

At the risk of sounding like that prize arse Simon Cowell, I have to say that Shoshannah owned that stage. She was confident, articulate, funny and resourceful, and one of the best speakers of the weekend. I am sure that we shall be seeing more of her on Weird Weekend stages in the future.

Lars Thomas was particularly interesting as he gave a remarkable insight into how and what the CFZ laboratory actually does. In fact, I have to admit that in the fifteen years or so that he has been running the CFZ laboratory, this was the first time that I had actually seen it, albeit in the form of a jpg projected onto a white sheet.

For the second year running the legendary Judge Smith, original drummer for the prog rock band Van Der Graff Generator, did a presentation based upon one of the books in his trilogy about the philosophy and science of life after death. I do not necessarily agree with his thesis, but it is a well-argued and intriguing one. I think, however, the thing that I find most interesting about it, is that it is the only cogent argument on the subject that I have ever read that leaves religion completely out of the equation. Interesting stuff.

The Saturday night was, as is customary, the time for Richard Freeman to present his latest CFZ expedition report, this time describing his October/November visit with CFZ Australia to Tasmania in search of the thylacine. As it is described by him in more detail elsewhere in this issue, I shall not go into any more detail, but sufficient to say it was all that we could have wanted.

Each Weird Weekend since 2003 we have held the annual CFZ Awards, during which we celebrate the doings of the great and the good of Forteana. This year, perhaps the most important of the recipients was our intern for much of 2013/4 Saskia England.

I also completely broke with tradition by giving her a cash award of fifty quid to help towards her educational wildlife trip to South Africa which will be taking place in March this year. She is an excellent young lady, and I am very proud to be able to announce that she has been accepted by all three of the universities to which she has applied to read Marine Biology. She is a lovely girl with a great future ahead of her, and if there is anything I or the CFZ in general can do in order to help make that future happen she only has to ask.

For the first time since 2002 when I reunited my old band The Amphibians from Outer Space, and for the first time since 1996 played in front of a live audience, we had music at the Weird Weekend.

Miss Crystal Grenade, has been described as Shakespeare's Sister fighting Amanda Palmer and Tori Amos in a dimly lit Victorian pub. With hand deformities. And that's a pretty good way of putting it. Miss Crystal Grenade explores the darker side of existence, framed by the concept of Freak Show life of the late 1800s. Whether she is indeed a late Victorian sideshow attraction who has only survived this long because of her fragile beauty and low cunning, or whether she is a character created by critically acclaimed singer and pianist Carol Hodge is a matter for discussion.

As I have written on many occasions, philosophically I see very little difference between what I do musically and within Forteana, and I have always wanted to have music, and indeed theatre, at the Weird Weekend, but I didn't know how other people would feel, so as Miss Crystal Grenade and her partner Pete the guitarist came on stage my fingers were well and truly crossed. But she was magnificent and made a lot of new fans. Music will definitely be a big part of the Weird Weekend from now on.

The new-look Weird Weekend was a definite success. We didn't have as many people as usual there, but that was at least partly the result of a conscious decision made by Andrea and myself, once we decided to move the gig to the new venue. We wanted to make sure that it worked in the new setting, and with the stylistic and philosophical changes that we had taken onboard before making any big effort at publicising the event. The school made about £1,200 profit out of the event and we made a couple of hundred quid, which is far less than usual, but - most importantly - we proved the new concept, and put the whole thing together in a very few weeks, when all sorts of other stuff, that I have no intention of going into here, was going on.

The publicity for the 2015 event starts here, and I think that it is going to be a bigger and even more successful event than last year, and will, I hope, make quite a lot of money both for us, and for this remarkable little school.

Letters

The editor and his compadres welcome letters for publication on all subjects covered by this magazine. However, we would like to stress that neither this magazine, or the CFZ are responsible for opinions expressed, which are purely those of the letter writer.

Take the snake to the Lake

Dear Jonathan,

I trust CFZ is thriving and you are both well.

When on a recce in Assam last week, I had an interesting encounter with a rather large King Cobra (photos attached).

Although they are not endangered, I thought you would like to see the fellow. We estimated it at 15 feet but no one cared to measure him.

I'll be returning there in March and have an expedition in Costa Rica in July.

John Blashford-Snell

I've just seen a face I can't forget

Hi Jon

In early June 2014 I was browsing through the Net and came across the image.

The image, taken as a photograph at the Museum of Traditional Medicine in Ho Chi Minh City, shows what looks like a complex hybrid between a man (the face), a snail, (the back) a porcupine, or a porcupine missing many quills, a squirrel (the tail) and the legs of a cloven hoofed animal.

Well this object fascinates me and in late June I emailed the Museum but my message

bounced back.

So I wrote to them and as of today (July 8th) I haven`t heard back. But I posted the image on Facebook on June 7th and the following conversation took place up to June 9th:

- Myself to Dale Drinnon (D.D.) Dale, any idea about this?
- D.D : That would be probably of Chinese origin and not depicting a Vietnamese " Native" Cryptid. And I suspect it is another depiction of a big cat from a garbled description, such as the Mantichora is supposed to be.
- Markus Buhler (M.B) : Well, it has horns and cloven hooves, a hump and quite un-feline tail. Seems much more probable to be some sort of bovine chimera.
- Dale Drinnon: Except that I said " Based on a garbled description…like a Mantichora (Manticore, Martichora) it could well have horns and cloven hooves, or bat wings even, and still be presumably   only   based   on   a misunderstood   description   of   an ordinary tiger
- Richard Muirhead: I emailed the Museum to ask them for more info but the email bounced back, so if I can find a postal address I`ll try that, or a forum for visitors to Vietnam.
- Markus Buhler: What I find really interesting is that the face looks quite European.
- Richard Muirhead: Do you mean the

face? If it is of Chinese origin (why Dale?) then it may have been influenced by the Colonial era, I will only know if I can get a message to the Museum & a reply.

Richard Muirhead
Macclesfield

Influencing His Owliness?

Good evening. I recently finished your book, Owlman, and today when researching something completely unrelated I found Louis Breton's image of Andras in that Samuel Weiser edition of The Goetia. I don't recall mention of Andras in Owlman, though perhaps I missed it. I'm sure I'm not the first person to point this out, but thought I'd email.

Best wishes,

Luke.

Marten cats in the enchanted woods

Dear Jon,

I understand that Richard Muirhead has found you a reference to a Devon Pine Marten today.

DEVON FALCONER

Fox-Fur Farm Visit By Plymouth Field Club

A visit to Mr. H. G. Hurrell's fox-fur farm. near Wrangaton, was paid by members of Plymouth and District Field Club, who were shown among other creatures pine martens and silver foxes. Mr. Hurrell explained that one or two martens, which had gone wild in his woods, were existing largely on insects. though only certain kinds were eaten.

Among other activities, Mr. Hurrell is a falconer, and showed a keen-eyed goshawk, trained to the chase, and capable of catching rabbits. This bird had taken its quota of rabbits, thereby assisting in their control and adding to the domestic larder.

Close to the house a nesting-box, full of fledgling great tits, was seen. The birds were fully coloured and ready for flight. Nesting holes of the nuthatch and tawny owl were inspected. also some young sparrow-hawks.

By lucky chance a greater horseshoe bat, of about 12in. wingspan, was on show. and a possibly fresh record for the locality was a water shrew, picked up in the road.

Botanists of the party, collecting on the way to Glazemeet, found bog pennywort. sundew, the pink-flowered bog pimpernel, and pale blue bell-flower.

THE PINE-MARTEN

That set me to checking the newspaper archive to see what I could come up with re that species in Devon.

I found the attached cutting from *The Western Morning News,* 22nd June 1947, which may be of interest.

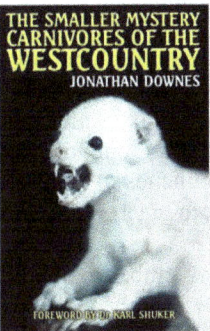

The report concerns a mention of pine martens from a fur farm at Wrangaton. The owner of the fur farm was Mr H.G. Hurrell, (who I note, Jon, you mention in your book *The Smaller Mystery Carnivores of the Westcountry*)

.
Members of the Plymouth Field Club visiting his fur farm were shown "among other animals pine martens and silver foxes."

Mr Hurrell is quoted as speaking of: "one or two martens, which had gone wild in the woods, were existing largely on insects, though only certain kinds were eaten."

This snippet may add some data to the controversy regarding releases/escapes of martens from Mr Hurrell's care.

Best regards

Bob Skinner

Reviews

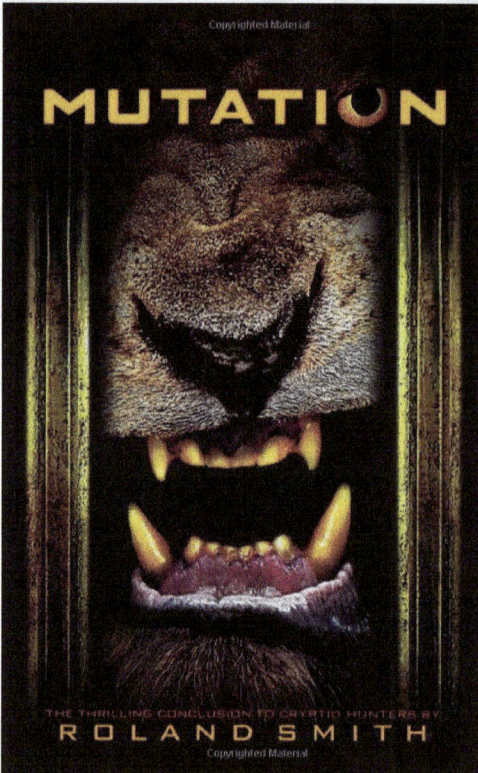

Hardcover: 352 pages
Publisher: Scholastic Press (30 Sep 2014)
Language: English

ISBN-10: 0545081807
ISBN-13: 978-0545081801

Seven years ago Corinna and I got married, and amongst our wedding presents was a bundle of books by a guy called Roland Smith. A few of years before, I had been on a research trip in the American heartland, in Illinois where I researched black panther sightings (no, I don't mean the political group), I filmed the seventeen year emergence of swarms of cicadas, and visited the locations where the Mad Gasser of Mattoon preyed upon his victims in the autumn of 1944.

Whilst I was in America I met up with someone who I had only previously contacted by email – a charming young lady called Elizabeth Clem, and it was she that bought us the Roland Smith books as a wedding present.

The one that I enjoyed most was one called *Cryptid Hunters* which told the story of thirteen year old Marty and his sister Grace who are whisked away to live with their uncle – an enormous bearded cripple who spends his life looking for unknown species of animals. Well, as I suspect that you might have gathered, I immediately felt an empathy with this chap, and avidly read the book.

By the end of the book we understand how Wolfe lost his leg, that Grace and Marty aren't really brother and sister at all, and we have been introduced to one of the most satisfyingly nasty bunches of villains that you could ever hope to meet.

I then realised that, to my embarrassment, that Roland Smith is a member of the Centre for Fortean Zoology , so I sent him an email congratulating him on a fine piece of writing. An hour or so later I got a charming email in reply, in which Roland admitted that both Wolfe and Marty had – in part – drawn inspiration from me, and more specifically from my 2004 autobiography *Monster Hunter*.

Ok, there are some glaring differences between me and my fictional counterparts. These are both material and philosophical. My leg was not bitten off by Mokele-mbembe in the depths of the Congo – injuries from a car crash, followed by diabetic neuropathy were what banjaxed my mobility, and although Wolfe made his money designing secret high tech stuff for the American government, my views about 'The Man' are, I believe, well known enough for me not to repeat them here.

It is, amusingly, not the first time that an analogue of me has been depicted in fiction; my friend and colleague Nick Redfern wrote a very funny book called *Three Men Seeking Monsters* which happily lampoons me and my mate Richard Freeman and I have turned up tangentially in at least two of his follow ups.

But this was different. It portrayed me as a hero rather than a camp alcoholic buffoon, and I was very pleased when Roland informed me that this was the first of a short series of books.

A few years later came a book called *Tentacles* in which our intrepid heroes are in the deep and hazardous waters near New Zealand trying to catch a giant squid.

Last year cane episode three – entitled *Chupacabra*, and much to my delight, a pre-release copy of the fourth instalment

arrived on my doormat a few days before the Weird Weekend. I read it in two sittings and I can wholeheartedly recommend it to anybody who likes immaculately crafted and well-written thrillers. This time the action takes place deep in the Amazon jungle as our heroes and villains square up for one last god almighty showdown.

The tone is a little darker this time around with allusions to Nazi genetic experimentation and more gruesome deaths than one usually finds in books written for the younger generation. But this is no bad thing. Roland Smith is a conservationist and ex-zookeeper who has the honour of being part of the United States Fish and Wildlife Services Red Wolf Recovery Team.

The story of the red wolf is a fascinating and intriguing one, and as a conservationist, I believe he has a duty not to sugar-coat his environmental message just because his target audience are in their teens.

Something that I have always admired about Smith is that he can write books about this particular demographic (a difficult undertaking at the best of times) and do it in a way that is neither condescending nor patronizing. When one considers that his writings do contain some quite abstruse concepts, then this achievement is even more impressive.

My only problem with this book is that it is the last in the series, which is a great pity, because I wasn't to follow Grace, Marty and their friends as they grow up, and I'm sure that the team have a lot more adventures left in them. JD

Adventures with reptiles and amphibians

Cold Blood

Richard Kerridge

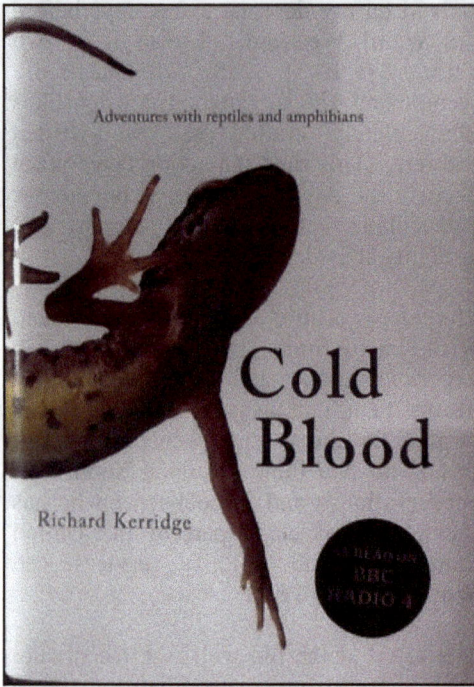

Hardcover: 304 pages
Publisher: Chatto & Windus (8 May 2014)
Language: Unknown
ISBN-10: 0701187956
ISBN-13: 978-0701187958
Product Dimensions: 22 x 16.4 x 4 cm

When I was an eleven-year-old boy in Hong Kong back in 1970, my mother used to play tennis with a lady called Mrs Muirhead and when, in the summer of that year, I was in hospital for some fairly serious surgery the aforementioned Mrs Muirhead brought her young son Richard, who was six at the time, round to visit me on my bed of pain. Fast forward forty-four years and Richard and I are still friends.

He is very good at buying Christmas and birthday presents, and knowing my fondness for things which hop and crawl, this year he brought me a book by someone called Richard Kerridge. The book is called 'Cold Blood: Adventures with Reptiles and Amphibians.'

This book actually recounts a fascinating journey. And it is a journey which I, too, travelled. It tells the story of a young man born into a relatively privileged background whose interest in the natural world took him out of that background and eventually to a place within the permissive further education, complete with cannabis and free love, of the time.

These days, one would imagine that the alpha male of an upper middle class family would be rather pleased that his son and heir was interested in the natural world. After all it beats the hell out of vandalizing telephone boxes, sneaking off to have a joint before school or getting drunk on cheap cider in the woods with a bevy of ne'er-do-well companions. But for Kerridge's father, every moment spent hunting for newts, catching lizards, or caring for a smooth snake (then and now the rarest British reptile) was time spent away from his studies, and another nail in the coffin of any hope of his son attaining the level of genteel respectability which he, as the master of the household, felt was appropriate.

I had exactly the same. And, like Kerridge, the fact that one of my chosen companions in hunting the byways of the local landscape after various creepy crawlies was the son of an unwed mother who had failed his eleven plus and went to the local Secondary Modern school propelled my father into the realms of apoplexy.

Both Kerridge's and my father did their best to discourage our activities because they felt that – in Kerridge's words – if we consorted with people who ate Kraft processed cheese triangles, it would somehow pollute our families' rarefied DNA and propel us towards social vulgarity.

Such class conscious bullshit is hard to countenance these days and I hope that, like capital punishment and the other extinct horrors listed above, it is now mercifully gone forever. But, I suspect not.

Class conscious snobbery (and remember it works both ways) has been one of the less attractive aspects of the human race for thousands of years.

Reading Kerridge's tale struck a very sympathetic chord for me and actually reopened some old wounds which I had not only closed for good but had actually forgotten about.

But the most extraordinary thing about this book is the realisation what Kerridge and his friends did in the 1960s, and – indeed – what I and my friends did in the 1970s, would be completely illegal today.

Under the provisions of the Wildlife and Countryside Act it is illegal to capture or keep species of most, if not all, British amphibians and reptiles. And it is this which opens up a whole slew of moral conundra.

As a society we are becoming increasingly divorced from the reality of the natural world, and I truly believe that this is a direct result of the fact that the last few generations of children have *not* been encouraged to pursue Natural History as a hobby.

The excuse that is given is always the big

C word - conservation, but the truth is that only one species in Britain has become extinct due to the predations of naturalists and collectors; the large copper butterfly in 1864.

And even this was compounded by the environmental changes wreaked by large scale drainage of its native fens.

The vast majority of species in the United Kingdom, if not the world, that have become extinct have done so due to environmental or social factors which have nothing at all to do with schoolchildren collecting caterpillars or butterflies or newts and keeping them in jam jars on their windowsills.

But unfortunately the legislation which should be used to prevent the increasing urbanisation of our countryside, and to stop fields and hedgerows and woodlands being grubbed up in the name of 'affordable housing' or leisure centres or motorway bypasses is being used counterproductively, and if we are not careful we shall become so divorced from the natural world that soon there will be entire generations who don't care enough about it to fight to save it.

Between the mid-19th Century and the mid-20th Century the study of Natural History was the most popular pastime in Britain amongst all social classes from Dukes to chimney sweeps.

Now it is seen as a peculiar eccentricity at best and at worst is demonised to the extent that a desire to study animals in your own home is seen as socially beyond the pale; well on its way to becoming as abhorrent as sexual abuse or terrorism.

Something has gone terribly wrong, and I truly believe that we are the last generation who shall be able to do anything about it! If we don't act now the only interaction future generations will have with the natural world is looking at it on television, or seeing cutesy creatures on increasingly fatuous video games. JD

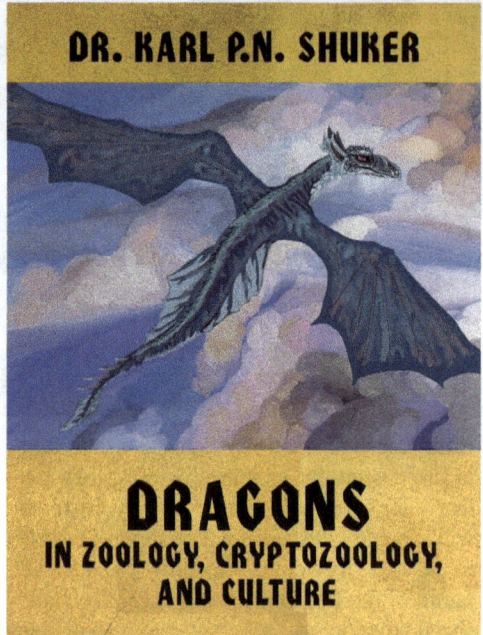

DR. KARL P.N. SHUKER

DRAGONS
IN ZOOLOGY, CRYPTOZOOLOGY, AND CULTURE

Hardcover: 220 pages
Publisher:: Coachwhip
Publications (18 Nov. 2013)
Language: English
ISBN-10: 1616462159
ISBN-13: 978-1616462154

The dragon is the great, great granddaddy of all monsters. It is found in cave paintings some 25,000 years old and manifests itself in the legends of every single culture on earth. Forget about demons, vampires, werewolves and zombies, the dragon is more ancient, widespread and powerful than all of them. Interestingly modern day sightings of such creatures persist (especial in Asia) despite most of mankind having relegated them to the realm of myth. Perhaps we have been far too hasty in our arrogant dismissal of the king of the monsters.

My fascination with dragons runs as far back as I can remember. I have written two books on the possible existence of real life dragons and hunted such beasts in West Africa and Indo-china.

Years ago my old friend Dr Karl Shuker wrote a book on dragons that I must confess I found a trifle light, more like a coffee table edition than the thick scholarly tomes he had penned before. Karl told me that it was the publishers that had wanted the book to be such and he had wanted to write something much weightier. Well that book is here at last and it was worth the wait.

Karl's book is separated into two halves. The first looks at the various types of dragon starting with lesser relations of dragons such as the limbless crawling worms, death dealing basilisks and the two legged wyvern (often substituted for true dragons by lazy and stupid film makers). It then moves on to the true western dragons marked by their huge size and by having four legs and two wings and the god-like rain-bringing dragons of the Orient.

There are many obscure types of dragon covered here including *Alklha*, a gigantic black dragon whose wings blot out the sun from the lore of Siberia's Buryat people, the *Dragua* of Albania that can become invisible, lives for a thousand years and is unkillable, the yellow Arminian *Svara* with its huge venomous fangs, the Hungarian giant, swamp-dwelling snake known as the *Zomok* that developed into the bat-winged tongue-twisting *Sarakanykigyo,* and the *Balaur* of Romania whose saliva hardened into precious gems.

Then the origins of dragon legends are examined. The bones of dinosaurs and other large prehistoric animals are put forward as one possible influence. In China dragons were supposed to shed and re-grow their bones along with their skin. Dinosaur bones are still called dragon bones in China.

Outsized reptiles such as giant crocodiles, lizards and constricting snakes encountered by early explorers could well have engendered legends of dragons.

These include the gigantic monitor lizard of Pleistocene Australia *Megalania prisca.* Moving on, Karl looks at reptilian cryptids such as sea serpents, the Alpine Tatzelworm and crested serpents in Africa. The second half of the book deals with the cultural impact of dragons. Art from ancient times and from all across the globe is looked at, from pottery and figures in Neolithic China, to the dragon of the Ishtar Gat in Babylon, and the serpent demon *Apep* of Egypt.

Renaissance paintings and their often tiny and ridiculous looking dragons (variously resembling rabid geese or winged dogs) were representations of Christianity defeating paganism. A far cry from the god-like beasts of earlier times.

Dragons' impact on the silver screen is far less than it should be. Whereas there are countless, dull films about vampires and zombies, the king of the monsters only rarely appears. As mentioned before dragons are often relegated to wyverns. Few gems amongst them include Larry Cohen's *Q The Winged Serpent* wherein a Mexican dragon worship cult prey a dragon god back into existence with human sacrifice in modern day New York.

Far better than any big screen dragon is one shown in the BBC series *Merlin.* For once a true, four-legged, two-winged dragon Kilgharrah is superior to any film dragon and is far better than the dragons in the otherwise excellent *Game of Thrones.*

By far the best section in the second part of the book is that which deals with dragons in literature. Alongside the well-known dragons such as Smaug (recently bastardized by Peter Jackson's awful film) in J.R.R. Tolkien's timeless book *The Hobbit* and those in Ursula LeGuin's *Earthsea* trilogy, Karl drew my attention to works I had not hitherto heard of, a treat for a committed bibliophile such as I.

These include Andrzej Sapkowski's *Witcher* series and Steven Brust's *To Reign in Hell.*

Elsewhere dragon influence in music, religion, sculpture and fashion are covered. There is far too much in the book to do it real credit in a review. What is clear is that no other animal, living, extinct or legendary has had such a huge impact upon mankind. I've said it before and I'll say it again, I have always believed that dragons have a basis in fact. Dr Shuker's new book has confirmed that once more. RF

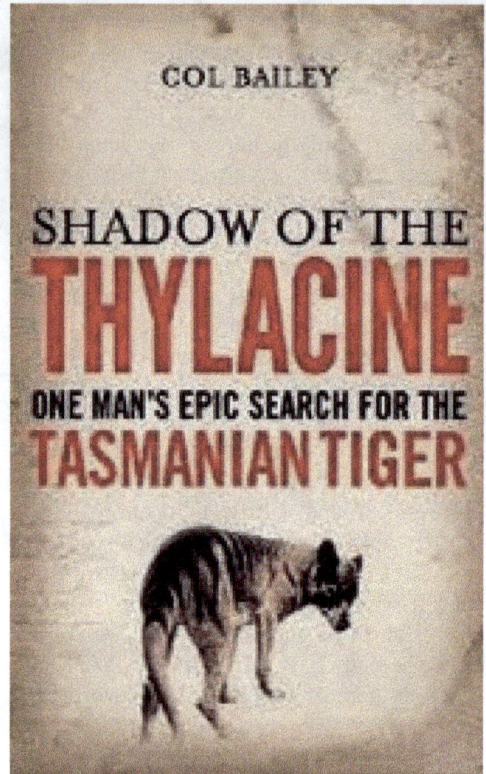

COL BAILEY

SHADOW OF THE

THYLACINE

ONE MAN'S EPIC SEARCH FOR THE

TASMANIAN TIGER

Format: Kindle Edition
File Size: 8839 KB
Print Length: 266 pages
Simultaneous Device Usage: Unlimited
Publisher: The Five Mile Press (29 April 2013)
ASIN: B00CL3N4R0

The thylacine, aka Tasmanian wolf / tiger is possibly the most iconic of all cryptids and certainly the most likely to exist. The flesh-eating marsupial has been described as the healthiest 'extinct' animal you will ever meet. However the greatest living champion of the thylacine is far less famous. He is a dedicated and modest man by the name of Colin (or Col) Bailey.

Rather than a book purely about the beast itself this is very much one man's story and what a fascinating story it is. Col, born on mainland Australia, was fascinated by wildlife from an early age. He had his ambition of becoming a vet quashed by circumstance, but continued his research into Australian wildlife. Then in 1967 he had an encounter at Coorong Lagoon in South Australia. Whilst canoeing he observed a creature on the bank; a thylacine. The sighting lasted some time and sparked a lifelong quest. He took a job as a landscape gardener and researched the creature in his spare time. During this period he met a few old trappers and bushmen including Reg Trigg who had apparently semi-tamed a female thylacine he caught in 1929.

The book contains information on the creature and what little is known of its biology, but this data is available elsewhere. What makes this book so interesting is Col's own story and the stories of the old bushmen and new witnesses he uncovers along the way. The folk who captured thylacines back in the bounty days are now all dead and Col did a valuable job of collecting information from them before it was lost to the ages. Moving to Tasmania Col continued his search. Self-funded, alone and in some truly remote, dangerous and difficult terrain you have to admire the man. In 1995 he was rewarded with an excellent sighting of a thylacine in the Weld Valley in South West Tasmania.

Col also documents his experience with production companies wanting to make films about the thylacine. Some were better than others and I noted one that, back in the mid-1990s had ripped off my own research and sent someone else instead of me!

Despite his advancing years Col still sallies forth into the bush. I can tell you from experience, Tasmania is one of the toughest wildernesses in the world and you have to admire the man's dedication. If anyone deserves to prove the continued existence of Tasmania's most splendid animal, it is surely Col Bailey. RF

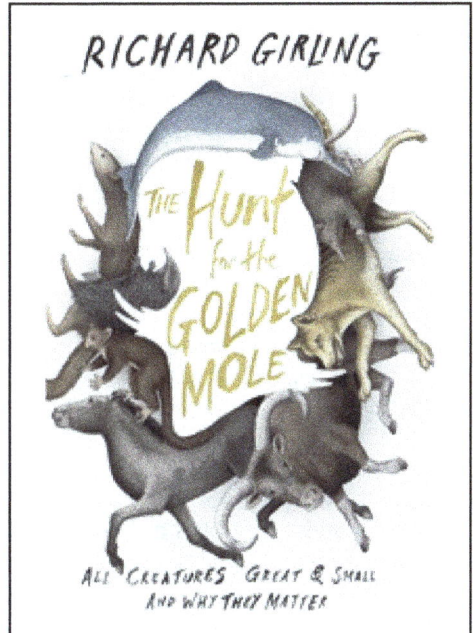

RICHARD GIRLING

THE Hunt for the GOLDEN MOLE

ALL CREATURES GREAT & SMALL AND WHY THEY MATTER

Hardcover: 320 pages
Publisher: Chatto & Windus (5 Jun. 2014)
Language: English
ISBN-10: 0701187158
ISBN-13: 978-0701187156

It is easy to fall into the trap of thinking that we live in a world where everything that comes out of mainstream outlets is dumbed down and aimed at the lowest common denominator. I know, I have fallen into that trap on a number of occasions. But books like this prove that paradigm wrong, and give hope for the future. It is the story of one man's hunt for that most elusive of all animals; a small creature only known from a few bones which make up the type specimen. It has never knowingly been seen alive, and no-one knows whether the animal still survives today or not, but as one of the major characters says, and as I have said on many occasions over the years, animals go extinct for a reason, and in this case there does not seem to be a reason for it to have become extinct.

Golden moles are small, insectivorous burrowing mammals native to southern Africa. They form the family Chrysochloridae. They are taxonomically distinct from the true moles, which they resemble, due to convergence. The golden moles bear a remarkable resemblance to the marsupial moles of Australia, so much so that, the marsupial/placental divide notwithstanding, arguments were once made that they were related, possibly because they are very primitive placentals and because of the many mole-like specializations.

The Somali golden mole was first described in 1964. It is mentioned in a number of textbooks, but the sole evidence for its existence is a tiny fragment of jawbone found in an owl pellet. The author discovers that this species is far from unique; the IUCN has many species which it calls 'Data Deficient'; species that we know next to nothing about. Just a brief look at the IUCN Red List website on my iPad as I sit here writing this, lists an incredible 1024 species of Data Deficient animals from the 2014 listing alone. This, as Max Blake pointed out in an address to the Weird Weekend some years ago, opens up a plethora of subjects more worthy of cryptozoological research than the things that seem to occupy the attention of an

embarrassing number of cryptozoologists. Dead racoons, and the bloated carcasses of dogs washed up on the morning tide do not a cryptid make, boys and girls. But I digress.

Richard Girling tells his enthralling tale against a background of the environmental genocide of the early 21st Century. It is tempting to think that we have progressed as a species over the past half century. When I used to read my Gran's copies of *Animals* magazine in the 1960s I was horrified to read how all the big animals of the planet would most likely have been hunted to extinction within a couple of decades. Half a century on we still have elephants, rhinos and the great whales so we like to fool ourselves that things are better. They are not. In most ways they are immeasurably worse! Girling explains the holocaust currently taking place against elephants and rhinos and it makes grim, nay, horrific reading.

As far as the hunt for the Somali golden mole is concerned, anyone who has ever tried to put cryptozoological investigations together will have realised that the cards were backed against Girling from the beginning, if only because Somalia is one of the most politically unstable places on the globe; a hotbed of piracy, terrorist training camps and God knows what other horrors. But he continues his quest, and reaches a satisfactory denouement. I won't go into any more details so I don't risk disappointing any potential readers. And you ALL should be potential readers. This book should be on the shelf of cryptozoologist, naturalist and conservationist alike. And, remember, every cryptozoologist should be a naturalist and a conservationist as well, or else the whole damn game isn't worth the candle.

An excellent book. JD

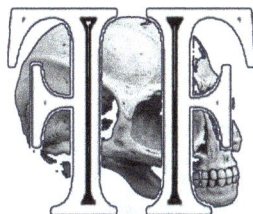

THE WORLD'S WEIRDEST PUBLISHING GROUP

We publish a lot of books. Indeed, I think that we could quite easily claim to be the world's foremost publishers of books about Fortean Zoology and allied disciplines. However, I feel that it would be unethical to review our own titles. So here, to end this edition of *Animals & Men*, is a brief look at some of the books we have put out in the last year.

We published the following books:

- *Hoaxed!* By Michael Newton (Fortean Words)
- *Wyrd* by J G Montgomery (Fortean Words)
- *The Magical Adventures of Henry Owl* (Fortean Fiction)
- *Judex Book Two - The Vibrating Spirit* by Judge Smith (Fortean Words)
- *Manbeasts* by Adam Davies (CFZ Press)
- *The Menagerie of Marvels* by Dr Karl Shuker (CFZ Press)
- *Atco and Grass* by Jim Jackson (CFZ Communications)
- *The Museum of the Future and Other Stories* by Dr Andrew May (Fortean Fiction)

You will be pleased to hear, I am sure, that Volume Two of George Eberhart's monumental *Mysterious Creatures* is nearly complete, and is with the author now for indexing. This has been a monumentally arduous endeavour, but I am glad that we took it on, and even more glad that we have nearly finished.

Ronan Coghlan has taken over publicising the books on Facebook and elsewhere, and I would like to publicly thank him for all his hard work here, and on the CFZ America blog which was opened this year.

I would also like to thank my current list of employees and interns: Andrea, Tammy, Jess, Danny and the two interns who completed their placements this year: Sheri and Saskia.

Thank you for all you have done my dears, it really is gratefully received. I would also like to thank Dr Andrew May for writing the Words from the Wild Frontier column twice a week, and commencing work on ebooks.

weird weekend **2015**

weekend

The Small School, Hartland, North Devon
www.cfz.org.uk

August 14-16 2015
TEL: +44 (0) 1237 431413

Three days of Monsters and Mysteries

For the second year running......

HARTLAND, YOU'VE NEVER HAD IT SO WEIRD